Lectures on
VON NEUMANN ALGEBRAS

DAVID M. TOPPING
Tulane University, Louisiana, U.S.A.

VAN NOSTRAND REINHOLD COMPANY
LONDON

NEW YORK CINCINNATI TORONTO MELBOURNE

VAN NOSTRAND REINHOLD COMPANY
Windsor House, 46 Victoria Street, London, S.E.1

INTERNATIONAL OFFICES
New York   Cincinnati   Toronto   Melbourne

Copyright © 1971 D. M. Topping

All rights reserved. No part of this publication may be reproduced, stored in a retrieval system, or transmitted by any means, electronic, mechanical, photocopying, recording or otherwise, without the prior permission of the copyright owner

Library of Congress Catalog Card No. 71-160203
ISBN 0 442 08563 X

First Published 1971

Printed in Great Britain by
Butler and Tanner Ltd
Frome and London

Contents

		Page
Preface		v
0	Terminology and Prerequisites	1
1	Monotone Convergence of Operators	7
2	von Neumann Algebras	11
3	The Geometry of Projections	17
4	Classifications of von Neumann Algebras	23
5	The Commutant and Type I	29
6	Type I Structure Theory	39
7	Some General Structure Theory	45
8	Topological Properties of *-Isomorphisms	51
9	Algebras on Separable Spaces	61
10	Generators	67
11	Hyperfinite Algebras	81
Appendix A.	Direct Sums and Matrix Algebras	95
Appendix B.	The Coupling Theorem	101
Appendix C..	Semifiniteness and the Commutant	107
References		111

Preface

These lectures had their formal origins in a graduate seminar at Tulane University in the Spring of 1967. During the academic year 1967, two advanced graduate courses were offered, one by Jacques Dixmier on C^*-algebras, and one by John Ernest on group representations. Fred Wright asked me to give a short series of lectures with the aim of providing additional background material for students in the aforementioned courses. My object was to pursue the most direct route to the theory of type I von Neumann algebras. In the Spring of 1968, I gave an amplified version of the original lectures in a seminar course at Indiana University. As questions about generators were very much in the air at this time, Peter Fillmore urged me to include some material on the subject. In the course of the lectures, Warren Wogen, one of my students, gave an elegant and elementary solution of the single generator problem for properly infinite algebras, and I have included his proof in §10.

There are many notable omissions, e.g., an example of a type III factor, existence of the trace in finite algebras, and reduction theory. I have avoided specific mention of the 'ultrastrong' and 'ultraweak' topologies. To make amends for these shortcomings, I have included Appendices B and C to push the continuous spatial theory a little further. Beyond this, of course, one can consult the excellent treatises of Dixmier [1], Kaplansky [10], and Schwartz [21].

These notes are intended to serve as a quick introduction to the subject. I suppose the reader to have some knowledge of single operators on Hilbert space, up through, say, the Spectral Theorem for bounded self-adjoint operators. Thus, to the reader who has a nodding acquaintance with 'operators, one at a time,' I have endeavored to present a sampler of 'operators in bunches.'

I have not been fastidious in providing references and credits for proofs, but have simply included the best proofs I know. Much of the theory presented here I learned from such men as Jacques Dixmier, Henry Dye, James Glimm, Ernest Griffin, Richard Kadison, Irving Kaplansky, Carl Pearcy, Arlan Ramsay, Teishirô Saitô, Irving Segal, and Fred Wright, through conversations writings and lectures.

I would like to thank my student, John Bunce, for proofreading the entire manuscript and making a number of helpful suggestions. I am grateful to the Ford Foundation and to the National Science Foundation for their support during various stages of the writing.

New Orleans, Louisiana, U.S.A.   D. M. T.

§0. Terminology and Prerequisites

In this section, we set down a list of facts with which we assume the reader is familiar. A few of these lack adequate references in the literature, and we have phrased them as exercises with connecting hints where possible (some references will be found in the bibliography). This material is basic to all of operator theory, and not just to the study of von Neumann algebras.

All operators that we deal with will be bounded and will act on a complex Hilbert space $\mathcal{H}$, with inner product $(\cdot|\cdot)$. We use ran(A) and null(A) to denote the range and null spaces, respectively, of an operator A. We abbreviate 'self-adjoint' to 's.a.'.

(1) For a set $\mathcal{S}$ of vectors in $\mathcal{H}$, let $[\mathcal{S}]$ denote the (norm) closed complex linear span of $\mathcal{S}$ in $\mathcal{H}$. For a family $\{\mathcal{E}_i\}$ of closed subspaces of $\mathcal{H}$, let $\vee \mathcal{E}_i$ denote the closed subspace $[\cup \mathcal{E}_i]$. Then $(\vee \mathcal{E}_i)^\perp = \cap \mathcal{E}_i^\perp$. If E_i is the (orthogonal) projection onto $\mathcal{E}_i$, then $E = \text{LUB } E_i$ exists as a projection and ran$(E) = \vee \mathcal{E}_i$. Further, ran$(I - E_i) = \mathcal{E}_i^\perp$, so $I - E = \text{GLB}(I - E_i)$. [7, §30.]

(2) For any operator A, null$(A^*)^\perp = [\text{ran}(A)]$.

(3) For any two operators A and B, null$(A^*A + B^*B)$ = null$(A) \cap$ null(B).

[Recall that an operator A is *positive* $(A \geq 0)$ if $(Ax|x) \geq 0$, for each vector $x \in \mathcal{H}$].

(4) Any positive operator A is s.a. (see (11) below) and has a unique positive square root S ($S^2 = A$) which is a norm limit of polynomials in A without constant terms [19; Theorem, p. 265]. (Note that for any operator T, $T^*T \geq 0$, and if S is s.a., $S^2 \geq 0$. If A is a s.a. operator and $|A| = (A^2)^{1/2}$, then $[\operatorname{ran}(A)] = [\operatorname{ran}(|A|)]$ (this follows from 2) and the equation $\operatorname{null}(A) = \operatorname{null}(|A|)$.)

(5) Two s.a. operators A and B have the same null space if and only if the same projections annihilate A and B.

(6) Let $\mathcal{E}$ be a (closed, linear) subspace of $\mathcal{H}$, T an operator on $\mathcal{H}$ and E the projection onto $\mathcal{E}$. Then: (a) $\mathcal{E}$ is *invariant* under T (i.e., $T\mathcal{E} \subset \mathcal{E}$) if and only if $ETE = TE$; and (b) $\mathcal{E}$ is invariant under T^* if and only if $\mathcal{E}^\perp$ is invariant under T. [7, Theorem 1, p. 45.]

(7) $\mathcal{E}$ *reduces* T (i.e., $\mathcal{E}$ and $\mathcal{E}^\perp$ are both invariant under T) if and only if $TE = ET$. [7, Theorem 2, p.45].

(8) $\mathcal{E}$ reduces T if and only if $\mathcal{E}$ is invariant under both T and T^*.

(9) If $A \geq 0$, then $(I + A)^{-1}$ exists.

(10) If A is s.a., then A has real spectrum (converse false). [7, Theorem 1, p. 54.]

(11) A is s.a. if and only if $(Ax|x)$ is real, for all $x \in \mathcal{H}$. [8, Theorem 4, p. 39.]

(12) $A \geq 0$ if and only if A is s.a. and has non-negative spectrum.

(13) If $A \geq 0$ and B is *any* operator, then $B^*AB \geq 0$.

(14) For Y s.a., $(I + iY)^{-1}$ exists. (*Hint.* Zero is not in the spectrum.)

(15) If $X \geq 0$ is invertible and Y is s.a., then $(X + iY)^{-1}$ exists. (*Hint.* use (14).)

(16) If $A, B \geq 0$ and $A^2 \leq B^2$, then $A \leq B$. In other words, if $0 \leq A \leq B$, then $A^{1/2} \leq B^{1/2}$ (converse false). (*Hints.* Let $\lambda > 0$

and let X and Y denote the real and imaginary parts, respectively, of $(\lambda I + A + B)(\lambda I + B - A)$. Then apply (9), (15) and (12.)

(17) In any ring with identity, existence of $(1 - xy)^{-1}$ implies existence of $(1 - yx)^{-1}$. (*Hint.* Expand $(1 - xy)^{-1}$ in a formal power series and guess the inverse of $1 - yx$.)

(18) For any two operators A and B, the spectra of AB and BA are the same, except perhaps for zero. (*Hint.* Use (17).)

(19) For $A, B \geqslant 0$ the spectrum of AB is non-negative, even though AB may fail to be s.a. (*Hint.* Use (18), write $B = B^{1/2}B^{1/2}$ and apply (13).)

(20) If $A, B \geqslant 0$ and if $AB = BA$, then $AB \geqslant 0$. [19; p. 265.]

(21) If $0 \leqslant A \leqslant B$ and if $AB = BA$, then $A^2 \leqslant B^2$. (*Hint.* Factor $B^2 - A^2$ and use (20).)

(22) If $0 \leqslant A \leqslant B$ and if A is invertible, so is B, and $0 \leqslant B^{-1} \leqslant A^{-1}$.

(23) The same operators commute with a positive ($\geqslant 0$) operator and its square (or square root). (*Hint.* Use (4).)

(24) For any s.a. operator A with spectrum $sp(A)$, $||A|| = r(A)$, where $r(A) = \sup\{|\lambda|: \lambda \in sp(A)\}$ is the *spectral radius* of A ([7; Theorem 2, p. 55]).

(25) Let $0 \leqslant A \leqslant E$ with E a projection. Then $A = AE = EA$. (*Hint.* The mapping $X \mapsto (I - E)X(I - E)$ preserves order by (13).)

(26) If $A \geqslant 0$ and E is the projection onto $[ran(A)]$, then $0 \leqslant A \leqslant ||A|| \cdot E$.

(27) For any operator T, $||T^*|| = ||T||$ and $||T^*T|| = ||T||^2$ ([7; Theorem 2, p. 39 and Theorem 4, p. 40]).

(28) For any s.a. operator A, $||A|| = \inf\{\alpha > 0: -\alpha I \leqslant A \leqslant \alpha \cdot I\}$ ([19; p.262]).

Let $\mathcal{B}(\mathcal{H})$ denote the ring of all operators on $\mathcal{H}$. We shall be concerned with three topologies on $\mathcal{B}(\mathcal{H})$.

I. *The Operator Norm (or Uniform) Topology.* Convergence in the *operator norm* of a net $\{T_n\}$ of operators to an operator T means that

$$||T - T_n|| \to 0 \text{ as } n \uparrow,$$

where $||T|| = \sup\{||Tx||: ||x|| = 1\}$.

II. *The Strong Operator Topology.* We say that the net $\{T_n\}$ of operators converges *strongly* (or in the *strong operator topology*) to the operator T if

$$||(T - T_n)x|| \to 0 \text{ as } n \uparrow,$$

for each vector $x \in \mathcal{H}$.

The strong operator topology on $\mathcal{B}(\mathcal{H})$ is the weakest (i.e., smallest) topology for which the mappings $T \mapsto Tx (x \in \mathcal{H})$ are continuous, when $\mathcal{H}$ is given its norm topology.

The strong operator topology is locally convex and is defined by the pseudonorms $\{p_x\}$, where

$$p_x(T) = ||Tx|| \qquad (x \in \mathcal{H}).$$

The strong operator topology is also a 'compact-open topology', where 'compact' refers to the discrete topology in $\mathcal{H}$, and 'open' refers to the norm topology in $\mathcal{H}$.

Basic strong operator neighborhoods $\mathcal{U}_s$ of an operator T are described by fixing $\epsilon > 0$, a finite set $\mathcal{F}$ of vectors in $\mathcal{H}$ and setting

$$\begin{aligned}\mathcal{U}_s &= \mathcal{U}_s(\mathcal{F}, \epsilon; T) \\ &= \{S \in \mathcal{B}(\mathcal{H}): ||(S - T)x|| < \epsilon, \, x \in \mathcal{F}\}.\end{aligned}$$

III. *The Weak Operator Topology.* We say that the net $\{T_n\}$ of operators converges *weakly* (or in the *weak operator topology*) to the operator T if

$$((T - T_n)x|y) \to 0 \quad \text{as } n\uparrow,$$

for all vectors $x, y \in \mathcal{H}$.

The weak operator topology on $\mathcal{B}(\mathcal{H})$ is the weakest (i.e., smallest) topology for which the mappings $T \mapsto (Tx|y)$ $(x, y \in \mathcal{H})$ are continuous.

The weak operator topology is locally convex and is defined by the pseudonorms $\{q_{x,y}\}$, where

$$q_{x,y}(T) = |(Tx|y)| \quad (x, y \in \mathcal{H}).$$

(By polarization [8, §71], the weak operator topology is also given by the pseudonorms $\{r_x\}$, where

$$r_x(T) = |(Tx|x)| \quad (x \in \mathcal{H}).)$$

The weak operator topology is also a 'compact-open topology', where 'compact' refers to the discrete topology in $\mathcal{H}$, and 'open' refers to the weak Hilbert space topology.

Basic weak operator neighborhoods $\mathcal{U}_w$ of an operator T are described by fixing $\epsilon > 0$, a finite set $\mathcal{F}$ of pairs of vectors in $\mathcal{H}$ and setting

$$\begin{aligned}\mathcal{U}_w &= \mathcal{U}_w(\mathcal{F}, \epsilon; T) \\ &= \{S \in \mathcal{B}(\mathcal{H}): |((S - T)x|y)| < \epsilon, (x, y) \in \mathcal{F}\}.\end{aligned}$$

We shall develop further properties of these topologies as we need them. For the present, let us merely remark that the following (proper) inclusions exist between these topologies:

$$\left\{\begin{array}{l}\text{weak} \\ \text{operator} \\ \text{topology}\end{array}\right\} \subset \left\{\begin{array}{l}\text{strong} \\ \text{operator} \\ \text{topology}\end{array}\right\} \subset \left\{\begin{array}{l}\text{operator} \\ \text{norm} \\ \text{topology}\end{array}\right\}.$$

(*Proof.* Suppose $||T - T_n|| \to 0$. Then $||(T - T_n)x|| \leq ||T - T_n||\cdot||x|| \to 0$ as $n\uparrow$. If $||(T - T_n)x|| \to 0$ for each vector $x \in \mathcal{H}$, as $n\uparrow$, then $|((T - T_n)x|y)| \leq ||(T - T_n)x||\cdot||y|| \to 0$, by the Schwarz Inequality).

§1. Monotone Convergence of Operators

We begin with a fundamental result which plays a role in operator theory parallel to that of the Monotone Convergence Theorem of Lebesgue in the theory of integration.

LEMMA 1. (Vigier [23]). *Let $\{A_n\}$ be a directed family of s.a. operators (i.e., given A_n, A_m there is an operator A_p in the family with A_n, $A_m \leq A_p$) which is bounded above by a s.a. operator B (i.e., $A_n \leq B$, for each n). Then $A = \text{LUB } A_n$ exists and $A_n \to A$ strongly.*

Proof. For m fixed, the family $\{A_n : A_m \leq A_n\}$ is cofinal in $\{A_n\}$, so we may assume that the original net is also bounded below. By translating and scaling, if necessary, we can further assume that $0 \leq A_n \leq I$, for all n.

The family $\{(A_n x | x)\}$ of real numbers is directed and bounded above by $||x||^2$. Let $F(x, x) = \text{LUB}(A_n x | x)$. Then $(A_n x | x) \to F(x, x)$. By polarization, $(A_n x | y) \to$ a limit $F(x, y)$. Note that the map $(x, y) \mapsto F(x, y)$ is conjugate bilinear, Hermitian symmetric and bounded, in fact by 1. Since the associated quadratic form $F(x, x)$ is real, there exists a unique s.a. operator A such that $(Ax | y) = F(x, y)$, for all $x, y \in \mathcal{H}$ (see [7, Theorem 1, p. 38]).

Clearly $A_n \to A$ weakly and $A_n \leq A$. If B is an upper bound for $\{A_n\}$, then $(Bx | x) \geq (A_n x | x)$, for all n, so that $(Bx | x) \geq \text{LUB}(A_n x | x) = (Ax | x)$ and $B \geq A$.

Finally, if $x \in \mathcal{H}$,

$$||(A - A_n)x||^2 \leq ||(A - A_n)^{1/2}||^2 \cdot ||(A - A_n)^{1/2}x||^2$$
$$\leq ((A - A_n)x|x) \to 0$$

since $0 \leq A - A_n \leq A \leq I$ implies $||(A - A_n)^{1/2}|| \leq 1$. Thus $A_n \to A$ strongly.

Recall that if $\{T_n\}$ is a net of operators and if $T_n \to T$ strongly, with $||T_n|| \leq 1$, for all n, then $T_n^2 \to T^2$ strongly (*Proof.* $T^2 - T_n^2 = (T - T_n)T + T_n(T - T_n)$, so $||(T^2 - T_n^2)x|| \leq ||(T - T_n)Tx|| + ||T_n|| \cdot ||(T - T_n)x|| \to 0$ as $n \uparrow$). This is no longer true if we drop the boundedness condition (see [9; Problem 91]).

LEMMA 2. *Let $0 \leq A \leq I$ be a s.a. operator and let E be the projection onto $[\operatorname{ran}(A)] = \operatorname{null}(A)^{\perp}$. Then the same projections annihilate A and E. Further, $0 \leq A \leq E$ and $AE = A = EA$; and E is the smallest projection with this property. Finally, E is the strong limit of a sequence of polynomials in A without constant terms.*

Proof. Let P be a projection with $PA = 0$. Then if $x \in \mathcal{H}$, $P(Ax) = 0$, so P vanishes on $[\operatorname{ran}(A)]$ and $PE = 0$. Conversely, if $PE = 0$, then for $x \in \mathcal{H}$, $P(Ax) = PE(Ax) = 0$, since $Ax \in \operatorname{ran}(E)$, so $PA = 0$.

In particular, $A(I - E) = 0 = (I - E)A$, so $AE = A = EA$. If P is a projection with $AP = A$ ($= PA$), then $A(I - P) = 0$ and $E(I - P) = 0$ or $E = EP \leq P$.

Also $E - A = E(I - A) \geq 0$ (see §0, 20)). Next note that if $0 \leq S \leq I$, then $S - S^2 = S^{1/2}(I - S)S^{1/2} \geq 0$ (see §0, 13)). In particular,

$$A^{1/2^{n+1}} - A^{1/2^n} = A^{1/2^{n+1}} - (A^{1/2^{n+1}})^2 \geq 0$$

so that $A \leq A^{1/2} \leq A^{1/4} \leq A^{1/8} \leq \ldots \leq E$. Let $G = \text{LUB } A^{1/2^n}$. Then $G = \text{strong-lim } A^{1/2^n}$ by Lemma 1. But $G^2 = \text{strong-lim}(A^{1/2^n})^2$ $= \text{strong-lim } A^{1/2^{n-1}} = G$ by the remark preceding the present lemma. Thus $A \leq G$ and $AG = A = GA$, so $E \leq G$. But clearly $G = \text{LUB } A^{1/2^n} \leq E$, so $E = G$.

Finally, $A^{1/2}$ (as well as successive square roots) can be uniformly approximated by a sequence of polynomials in A without constant terms, and the result follows.

THEOREM 1. *Let $\mathcal{S}$ be a strongly closed real linear space of s.a. operators such that $A^2 \in \mathcal{S}$ whenever $A \in \mathcal{S}$. Then $\mathcal{S}$ contains a largest projection E such that $AE = A = EA$, for each $A \in \mathcal{S}$. Further, $\mathcal{S}$ contains the projection onto the closure of the range of each of its operators, and also the (relative) spectral family of each of its operators. In particular, $\mathcal{S}$ is the norm-closed real linear span of its projections.*

Remark. $A^n \in \mathcal{S}$ if $A \in \mathcal{S}$, for $n = 2, 3, \ldots$, since $A^{n+1} = \frac{1}{2}[(A + A^n)^2 - A^2 - A^{2n}]$ (use induction).

Proof. First note that if $A \in \mathcal{S}$, then $|A| = (A^2)^{1/2} \in \mathcal{S}$. For $|A|$ is the norm limit of real polynomials in A^2 without constant terms. Also $A = A^+ - A^-$, where $A^\pm = \frac{1}{2}(|A| \pm A) \in \mathcal{S}$.

Now if A is any s.a. operator, then $[\text{ran}(A)] = [\text{ran}(|A|)]$ (see §0, 4)—more is true, namely $\text{ran}(T) = \text{ran}(TT^*)^{1/2}$, for any operator T, but we do not need this). Thus if $A \in \mathcal{S}$ and if E is the projection onto $[\text{ran}(A)] = [\text{ran}(|A|)]$, then $E \in \mathcal{S}$ by Lemma 2 (normalizing A if necessary).

For $A \in \mathcal{S}$, let E_A denote the projection onto $[\text{ran}(A)] = \text{null}(A)^\perp$ (see §0, 2)). For $A, B \in \mathcal{S}$, $\text{null}(A^2 + B^2) \subset \text{null}(A), \text{null}(B)$ (see §0, 3)), so $E_A, E_B \leq E_{A^2+B^2}$. Thus the family $\{E_A\}_{A \in \mathcal{S}}$ is directed and bounded above (by I) so $E = \text{LUB } E_A$ ($A \in \mathcal{S}$) exists, $E_A \to E$ strongly by Lemma 1, and $E \in \mathcal{S}$.

But E is the projection onto $(\cap \{\text{null}(A): A \in \mathcal{S}\})^\perp = \mathbf{V}\{\text{null}(A)^\perp: A \in \mathcal{S}\} = \mathbf{V}\{[\text{ran}(A)]: A \in \mathcal{S}\}$. Thus if $A \in \mathcal{S}$, $A(I - E)x = 0$, for all $x \in \mathcal{S}$, so $AE = A = EA$.

For $A \in \mathcal{S}$, the *relative spectral family* $\{E_\lambda\}$ *of* A is defined as follows. With λ real, the operator $(A - \lambda E)^+$ belongs to $\mathcal{S}$. Hence by Lemma 2, the projection F_λ onto $[\text{ran}(A - \lambda E)^+]$ is in $\mathcal{S}$, and hence $F_\lambda \leqslant E$. Define $E_\lambda = E - F_\lambda$. Then

(1) $E_\lambda \leqslant E_\mu$ if $\lambda \leqslant \mu$.

(2) $E_\lambda = \underset{\lambda < \mu}{\text{GLB }} E_\mu$.

(3) $\underset{\lambda < 0}{\text{GLB }} E_\lambda = 0$ and $\underset{\lambda > 0}{\text{LUB }} E_\lambda = E$.

(4) $A = \int_{-\infty}^\infty \lambda \, dE_\lambda$ and each $E_\lambda \in \mathcal{S}$.

(see the proof of the usual Spectral Theorem, e.g., in [19, pp. 279-280]). If $\{P_\lambda\}$ is the (usual) spectral family of the operator A, the relation with $\{E_\lambda\}$ is simple: $E_\lambda = EP_\lambda$.

§2. von Neumann Algebras

A *von Neumann algebra* is a (complex) ∗-algebra of operators on a Hilbert space, which is closed in the strong operator topology. Put differently, $\mathfrak{A}$ is a von Neumann algebra if $aA + B$, AB and A^* all belong to $\mathfrak{A}$ whenever A and B do, and a is a complex scalar, as do limits of strongly convergent nets of operators from $\mathfrak{A}$.

For a set $\mathcal{S}$ of operators on $\mathcal{H}$, the *commutant* of $\mathcal{S}$ is the set

$$\mathcal{S}' = \{T \in \mathcal{B}(\mathcal{H}): TS = ST, \text{ for all } S \in \mathcal{S}\}.$$

Observe that $\mathcal{S}'$ is closed in *all* operator topologies. For $\mathcal{S}' = \cap \{S\}' (S \in \mathcal{S})$ and $\{S\}'$ is the null space of the continuous function $T \mapsto TS - ST$. Also $\mathcal{S} \subset \mathcal{S}''$, $\mathcal{S}''' = \mathcal{S}'$ and $\mathcal{S}'$ is an algebra containing the identity operator I, but in general it need not be a ∗-algebra. For example if $\mathcal{S}$ consists of the single matrix

$$\begin{pmatrix} 0 & 0 \\ 1 & 0 \end{pmatrix}$$

then $\mathcal{S}'$ is the lower triangular algebra of matrices of the form

$$\begin{pmatrix} \alpha & 0 \\ \beta & \alpha \end{pmatrix}.$$

The next result imparts a special algebraic flavor to the subject.

THE DOUBLE COMMUTANT THEOREM (von Neumann [14]).
Let $\mathcal{A}$ be a von Neumann algebra with $I \in \mathcal{A}$. Then $\mathcal{A} = \mathcal{A}''$.

Proof. Take $A'' \in \mathcal{A}''$. For $x \in \mathcal{H}$, $\mathcal{A}x = \{Ax : A \in \mathcal{A}\}$ is a linear subspace and its closure $[\mathcal{A}x]$ is invariant under $\mathcal{A}$. Let E' be the projection onto $[\mathcal{A}x]$ and let $A \in \mathcal{A}$. By invariance (§0, 6)), $E'AE' = AE'$ and since $A^* \in \mathcal{A}$, $E'A^*E' = A^*E'$. Taking adjoints, we get $E'A = E'AE' = AE'$, so $E' \in \mathcal{A}'$. Hence E' commutes with every operator in $\mathcal{A}''$, so that $\mathcal{A}''$ leaves $[\mathcal{A}x]$ invariant. But $[\mathcal{A}x]$ contains x, since $I \in \mathcal{A}$. Hence $A''x \in [\mathcal{A}x]$ and $A''x = \lim A_n x$, for some sequence $\{A_n\}$ of operators in $\mathcal{A}$.

Now let $\tilde{\mathcal{H}} = \mathcal{H} \oplus \ldots \oplus \mathcal{H}$ (n copies) be the Hilbert space of all n-tuples $(x_1, \ldots, x_n)$ with $x_i \in \mathcal{H}$ and inner product given by

$$(\tilde{x}|\tilde{y}) = \sum_{i=1}^{n} (x_i|y_i),$$

where $\tilde{x} = (x_1, \ldots, x_n)$ and $\tilde{y} = (y_1, \ldots, y_n)$. For $B \in \mathcal{B}(\mathcal{H})$, let

$$\tilde{B} = \begin{pmatrix} B & & & 0 \\ & B & & \\ & & \cdot & \\ & & & \cdot \\ 0 & & & B \end{pmatrix}$$

($n \times n$ matrix with operator entries); $\tilde{B}$ acts on $\tilde{\mathcal{H}}$ as a bounded operator by

$$\tilde{B}\tilde{x} = (Bx_1, Bx_2, \ldots, Bx_n),$$

where $\tilde{x} = (x_1, x_2, \ldots, x_n) \in \tilde{\mathcal{H}}$. The mapping $B \to \tilde{B}$ is an injective *-algebra homomorphism of $\mathcal{B}(\mathcal{H})$ into $\mathcal{B}(\tilde{\mathcal{H}})$. Let $\tilde{\mathcal{A}} = \{\tilde{A} : A \in \mathcal{A}\}$. Clearly $\tilde{\mathcal{A}}$ is a *-algebra. We assert that $(\tilde{\mathcal{A}})' = \widetilde{(\mathcal{A}')}$, in other words, $(\tilde{\mathcal{A}})' = \{(A'_{ij}) : A'_{ij} \in \mathcal{A}', i, j = 1, 2, \ldots, n\}$. For

$0 = \tilde{A}(B_{ij}) - (B_{ij})\tilde{A} = (AB_{ij}) - (B_{ij}A) = (AB_{ij} - B_{ij}A)$ if and only if $AB_{ij} = B_{ij}A$, for all i and j. In particular, $\widetilde{A''}(A'_{ij}) = (A'_{ij})\widetilde{A''}$. Thus $A''\tilde{x} \in [\tilde{\mathfrak{A}}\tilde{x}]$ for each $\tilde{x} \in \tilde{\mathcal{H}}$ (note that $\tilde{I} \in \tilde{\mathfrak{A}}$ and argue as before for one vector $x \in \mathcal{H}$).

With $\tilde{x} = (x_1, \ldots, x_n)$, $\widetilde{A''}\tilde{x} = (A''x_1, \ldots, A''x_n)$ is the limit of a sequence $(A_n x_1, \ldots, A_n x_n)$ with each $A_n \in \mathfrak{A}$.

This shows that $\mathfrak{A}$ is strongly dense in $\mathfrak{A}''$, and equality follows.

Remark. We have actually shown that for a $*$-algebra $\mathfrak{A} \subset \mathcal{B}(\mathcal{H})$ with $I \in \mathfrak{A}$, the following conditions are equivalent: (1) $\mathfrak{A}$ is strongly closed; (2) $\mathfrak{A} = \mathfrak{A}''$; and (3) $\mathfrak{A}$ is weakly closed.

PROPOSITION 1. *Let $\mathfrak{A}$ be a von Neumann algebra which does not contain the identity operator I. Then $\mathfrak{A}$ contains a largest projection E satisfying $AE = A = EA$, for all $A \in \mathfrak{A}$. If $\mathfrak{M} = \{\lambda I\} + \mathfrak{A}$, then $\mathfrak{M} = \mathfrak{M}''$ is a von Neumann algebra containing $\mathfrak{A}$ as a maximal ideal. Further,*

$$\mathfrak{A} = \{M \in \mathfrak{M}: ME = M = EM\} = E\mathfrak{M}E.$$

Proof. Existence of E is immediate from Theorem 1. Strong closure of $\mathfrak{M}$ follows from linear topological space theory, but is also easy to see directly. Suppose $\{M_n\}$ is a net in $\mathfrak{M}$ with $M_n = \lambda_n I + A_n \to M$ strongly, where $A_n \in \mathfrak{A}$ and the λ_n's are complex scalars. Writing $E^- = I - E$, we have $\lambda_n E^\perp = M_n E^\perp \to M E^\perp$ strongly, so that $\{\lambda_n\}$ converges to some scalar λ. Hence $A_n = M_n - \lambda_n I \to M - \lambda I = A$ strongly, and $A \in \mathfrak{A}$ since $A_n \in \mathfrak{A}$ and $\mathfrak{A}$ is strongly closed. Thus $M = \lambda I + A \in \mathfrak{M}$.

Being a hyperplane and an ideal in $\mathfrak{M}$, $\mathfrak{A}$ is maximal. Finally, if $M \in \mathfrak{M}$ and $ME = M = EM$, then $M \in \mathfrak{A}$ by the ideal property, since $E \in \mathfrak{A}$.

From now on, when convenient, we shall assume that the projection E described in Proposition 1 is the identity operator I.

Exercise. If $E \neq I$, show that the restriction $\mathcal{A}|_{\text{ran}(E)}$ of $\mathcal{A}$ to the range of E is a von Neumann algebra on $\text{ran}(E)$ which is *-isomorphic to $\mathcal{A}$ (with $\mathcal{A}$ and E as in Proposition 1).

PROPOSITION 2. *Let $\mathcal{A}$ be a von Neumann algebra and let $\{E_n\}$ be any family of projections in $\mathcal{A}$. Then* LUB E_n *and* GLB E_n *(the spatial lub and glb — see §0, 1) belong to $\mathcal{A}$.*

Proof. If E and F are projections in $\mathcal{A}$, then $E \vee F$ (by definition, the projection onto $[\text{ran}(E) + \text{ran}(F)]$) is the projection onto $[\text{ran}(E + F)]$. For $\text{null}(E + F) = \text{null}(E) \cap \text{null}(F)$ (see §0, 3)) so that $[\text{ran}(E + F)] = \text{ran}(E) \vee \text{ran}(F)$. Hence $E \vee F \in \mathcal{A}$ if $E, F \in \mathcal{A}$. Also $E \wedge F = I - [(I - E) \vee (I - F)] \in \mathcal{A}$.

We can therefore enlarge the family $\{E_n\}$ by adjoining all finite *lub's* of the E_n's (without changing the *lub* of the family), and thereby assume that $\{E_n\}$ is a directed family. By Lemma 1, $E = \text{LUB } E_n$ is a strong limit of the E_n's and since everything is sight is bounded (≥ 0 and $\leq I$), $E^2 = E = E^* \in \mathcal{A}$ (see §1, remark before Lemma 2). As in the proof of Theorem 1, E is the projection onto $\vee \text{ran}(E_n)$. The argument for GLB E_n is dual (take finite *glb's*).

Remarks. (1) Let $\mathcal{A}$ be a von Neumann algebra and let $\mathcal{S} = \{A \in \mathcal{A}: A^* = A\}$. Then $\mathcal{S}$ satisfies the hypotheses of Theorem 1, and $\mathcal{A} = \mathcal{S} + i\mathcal{S}$. By the Spectral Theorem, if $S \in \mathcal{S}$ and if f is a bounded Baire function defined on the spectrum of S, then $f(S) \in \mathcal{S}$ (see [12; Theorem, p. 93]). (2) Given a von Neumann algebra and an operator T (on the same Hilbert space). To verify that $T \in \mathcal{A}'$ it is enough to check any one of the following:

(i) $TS = ST$, for each $S = S^* \in \mathcal{A}$ (by (1) above).

(ii) $TE = ET$, for each projection $E \in \mathcal{A}$ (by (1) above and Theorem 1).

(iii) $TU = UT$, for each unitary $U \in \mathcal{A}$ (To see this, note that each operator in $\mathcal{A}$ is a linear combination of two s.a. contractions. Hence we restrict our attention to $A \in \mathcal{A}$ with $A^* = A$ and $||A|| \leq 1$. Let $U = A + i(I - A^2)^{1/2}$. Then $U \in \mathcal{A}$ is unitary and $A = \frac{1}{2}(U + U^*)$. Thus each operator in $\mathcal{A}$ is a linear combination of four unitaries in $\mathcal{A}$.)

POLAR DECOMPOSITION. *Let $\mathcal{A}$ be a von Neumann algebra and let $A \in \mathcal{A}$. Then there is a partial isometry $U \in \mathcal{A}$ and a positive operator $H \in \mathcal{A}$ such that $A = UH$ and $\mathrm{null}(U)^{\perp} = [\mathrm{ran}(H)]$. Moreover, U and H are the unique operators with these properties* $(H = (A^*A^{1/2})$.

Proof. Define $U((A^*A)^{1/2}x) = Ax$ and note that $((A^*A)^{1/2}x | (A^*A)^{1/2}x) = (Ax|Ax)$. Thus U is isometric from $\mathrm{ran}((A^*A)^{1/2})$ to $\mathrm{ran}(A)$. Define U to be zero on $\mathrm{ran}((A^*A)^{1/2})^{\perp}$ and extend by continuity. Then U is a partial isometry with initial space $[\mathrm{ran}(A^*A)^{1/2}]$ and final space $[\mathrm{ran}(A)]$, and $A = U(A^*A)^{1/2}$. If $H \geq 0$ and $A = VH$, where V is a partial isometry with $\mathrm{null}(V)^{\perp} = [\mathrm{ran}(H)]$, then $A^*A = HV^*VH = H^2$ and $H = (A^*A)^{1/2} \in \mathcal{A}$. Also $V = U$.

To show $U \in \mathcal{A}$, it is enough, by the Double Commutant Theorem, to show that U commutes with each unitary $W \in \mathcal{A}'$ (since each operator in $\mathcal{A}'$ is a linear combination of four unitaries in $\mathcal{A}'$ by remark (2) (iii) above). Now W commutes with H, $A = W^*AW = (W^*UW)(W^*HW) = (W^*UW)H$ and W^*UW is a partial isometry. It is easy to see that $\mathrm{null}(W^*UW)^{\perp} = [\mathrm{ran}(W^*HW)] = [\mathrm{ran}(H)]$, so the uniqueness property of U implies $W^*UW = U$, and U commutes with W.

§3. The Geometry of Projections

The presence of projections in a von Neumann algebra (Theorem 1) provides a geometric tool for studying its structure. Projections, and the relations they inherit under partial isometry equivalence within the algebra itself, give rise (see §4) to several important classifications (discrete, continuous, finite and infinite) that are reminiscent of measure theory.

With regard to structure theory, the most powerful result, by far, in this section is the Comparison Theorem.

LEMMA 3. *For any two projections E and F in a von Neumann algebra $\mathcal{A}$, there is a symmetry (i.e., a s.a. unitary) $S \in \mathcal{A}$ such that $S(EF)S = FE$.*

Proof. Let $A = E + F - I$ and set $S = (P - Q) + R$, where P, Q and $I - R$ are the projections onto $[\operatorname{ran}(A^+)]$, $[\operatorname{ran}(A^-)]$ and $[\operatorname{ran}(A)]$, respectively. Then P, Q and R are mutually orthogonal, in (Theorem 1) and $P + Q + R = I$. Thus $S^2 = I$ and $A = S \cdot |A|$. Now E and F commute with $|A|^2 = A^2$ and since a positive operator and its square root commute with the same operators (§0, 23), E and F commute with $|A|$. Also $|A| = SA = SE = S(I - F)$ and hence $SEF = SAF = |A| \cdot F = F \cdot |A| = FES$ so that $S(EF)S = FE$.

LEMMA 4. *If E and F are two projections with $\mathcal{E} = \operatorname{ran}(E)$ and $\mathcal{F} = \operatorname{ran}(F)$, then*

(1) $[\operatorname{ran}(EF)] = \mathcal{E} \cap (\mathcal{E} \cap \mathcal{F}^\perp)^\perp$, and
(2) $[\operatorname{ran}(FE)] = \mathcal{F} \cap (\mathcal{F} \cap \mathcal{E}^\perp)^\perp$.

Proof. We first note that

$$\operatorname{null}(EF) = \mathcal{F}^\perp \oplus (\mathcal{F} \cap \mathcal{E}^\perp).$$

For if $EFz = 0$, $z = x + y$ with $x \in \mathcal{F}^\perp$ and $y \in \mathcal{F}$. But $0 = EFz = EFx + EFy = Ey$ (since $Fx = 0$ and $Fy = y$), so $y \in \mathcal{E}^\perp$ also. Thus $z \in \mathcal{F}^\perp \oplus (\mathcal{F} \cap \mathcal{E}^\perp)$.

Conversely, if $z = x + y$, $x \in \mathcal{F}^\perp$, $y \in \mathcal{F} \cap \mathcal{E}^\perp$, then $EFz = EFx + EFy = Ey = 0$ (again, $Fx = 0$ and $Fy = y$), so $z \in \operatorname{null}(EF)$.

Finally, $[\operatorname{ran}(EF)] = \operatorname{null}(FE)^\perp = (\mathcal{E}^\perp \oplus (\mathcal{E} \cap \mathcal{F}^\perp))^\perp = \mathcal{E} \cap (\mathcal{E} \cap \mathcal{F}^\perp)^\perp$. Interchange E and F to get (2).

For an operator T, let $\operatorname{rp}(T)$ denote the projection onto $[\operatorname{ran}(T)]$. We call $\operatorname{rp}(T)$ the *range projection* of T. For convenience, we write $E^\perp = I - E$ in what follows.

COROLLARY 1. *For projections E and F,*
(1) $\operatorname{rp}(EF) = E - E \wedge F^\perp$.
(2) $\operatorname{rp}(FE) = F - E^\perp \wedge F$.
(3) $\operatorname{rp}(EF^\perp) = E - E \wedge F$.
(4) $\operatorname{rp}(F^\perp E) = E \vee F - F$.

LEMMA 5. *For any two projections E and F, the projections $P = E - E \wedge F^\perp$ and $Q = F - E^\perp \wedge F$ satisfy the relations $P \wedge Q^\perp = 0 = P^\perp \wedge Q$.*

Proof. First note that $F = E \vee (F \wedge E^\perp)$ if $E \leq F$, and dually, $E = F \wedge (E \vee F^\perp)$ if $E \leq F$. By the second relation, $F^\perp = (E \vee F^\perp) \wedge (F^\perp \vee (E \vee F^\perp)^\perp)$. Hence
$E \wedge (E \wedge F^\perp)^\perp \wedge (F^\perp \vee (E^\perp \wedge F)) = E \wedge (E \vee F^\perp) \wedge (F^\perp \vee (E^\perp \wedge F)) \wedge (E \wedge F^\perp)^\perp = (E \wedge F^\perp) \wedge (E \wedge F^\perp)^\perp = 0$. Interchanging E and F gives the other equation.

COROLLARY 2. *For any two projections E and F, the projections $P = E \vee F - F$ and $Q = E - E \wedge F$ satisfy the relations $P \wedge Q^\perp = 0 = P^\perp \wedge Q$.*

Proof. In Lemma 5, replace F by $F^\perp$ and note that $F^\perp - E^\perp \wedge F^\perp = E \vee F - F$.

THEOREM 2. *Let E and F be projections in a von Neumann algebra $\mathfrak{A}$ such that $E \wedge F^\perp = 0 = E^\perp \wedge F$. Then there is a symmetry $S \in \mathfrak{A}$ with $SES = F$.*

Proof. By Corollary 1(1), $\mathrm{rp}(EF) = E$ and by (2), $\mathrm{rp}(FE) = F$. Let S be as in Lemma 3. Then $S \in \mathfrak{A}$ and $S(EF)S = FE$. For any vector $x \in \mathcal{H}$, $ESx \in [\mathrm{ran}(EF)]$, so there is a sequence $\{x_n\}$ of vectors in $\mathcal{H}$ with $ESx = \lim(EF)S^2 x_n$ (recall that $S^2 = I$). Thus $SESx = S \lim(EF)S^2 x_n = \lim(S(EF)S)Sx_n = \lim(FE)Sx_n \in [\mathrm{ran}(FE)] = \mathrm{ran}(F)$, so that $SES \leq F$. By symmetry, $SFS \leq E$. But $SES - F = S(E - SFS)S \geq 0$ (see §0, 13), so $F \leq SES$ and finally, $SES = F$.

Two projections E and F in a von Neumann algebra $\mathfrak{A}$ are said to be *equivalent*, written $E \sim F$, if there is a partial isometry $U \in \mathfrak{A}$ with $E = U^*U$ and $F = UU^*$. We write $E \leq F$ if $E \sim G \leq F$, for some projection $G \in \mathfrak{A}$.

COROLLARY 3. *For two projections E and F in a von Neumann algebra $\mathfrak{A}$,*

(1) $E \vee F - F \sim E - E \wedge F$ *(The Parallelogram Law).*
(2) $E - E \wedge F^\perp \sim F - E^\perp \wedge F$.

Proof. If there is a symmetry $S \in \mathfrak{A}$ such that $SPS = Q$, put $U = SP$. Then $U \in \mathfrak{A}$, $P = U^*U$ and $Q = UU^*$. Now apply Lemma 5 and Corollary 2 in conjunction with Theorem 2.

If we think of projections E and F in a von Neumann algebra as 'measurable sets' in a measure space, with $E \sim F$ meaning that E and F have the same 'measure', then the next result can be viewed as a symmetric formulation of complete additivity of the measure.

ADDITIVITY OF EQUIVALENCE. *Let $\{E_n\}$ and $\{F_n\}$ be orthogonal families of projections (indexed by the same set) in a von Neumann algebra $\mathfrak{A}$ and suppose that $E_n \sim F_n$, for each n. Then $E \sim F$, where $E = $ LUB E_n and $F = $ LUB F_n.*

Proof. Choose partial isometries $\{U_n\}$ with $U_n \in \mathfrak{A}$, $E_n = U_n^* U_n$ and $F_n = U_n U_n^*$. For each finite set φ of indices, let

$$U_\varphi = \Sigma U_n \qquad (n \in \varphi).$$

Since the collection $\Phi = \{\varphi\}$ of all such sets is directed by inclusion, the family $\{U_\varphi\}_{\varphi \in \Phi}$ is a net of operators in $\mathfrak{A}$. Now strong convergence of $\{U_\varphi\}$ means that each of the nets $\{U_\varphi x\}_{\varphi \in \Phi}$ converges in $\mathcal{H}$, for $x \in \mathcal{H}$. Since the $U_n x$'s are pairwise orthogonal, strong convergence is equivalent to finiteness of

$$\Sigma ||U_n x||^2 \left(= \lim_{\varphi \in \Phi} \sum_{n \in \varphi} ||U_n x||^2 \right).$$

Now $\Sigma ||U_n x|| = \Sigma(U_n^* U_n x | x) = \Sigma(E_n x | x) = (Ex|x) < \infty$. Replacing U_n by U_n^*, we see that ΣU_n^* is strongly convergent too. Let $U = $ strong-lim $\Sigma_{n \in \varphi} U_n$ ($\varphi \in \Phi$). Then convergence of ΣU^* is a fortiori weak, and since the operation of taking adjoints ($T \mapsto T^*$) is weak operator continuous, we have $U^* = \Sigma U_n^*$ (in the *strong* operator topology—if a net converges weakly *and* strongly, the limits agree).

Finally, $U_n^* U_m = (U_n^* F_n)(F_m U_m) = \delta_{nm} E_n$ so $U^* U = \Sigma U_n^* U_n = \Sigma E_n = E$. Replacing U_n by U_n^* gives $UU^* = F$.

LEMMA 6. *If $\mathfrak{A}$ is a von Neumann algebra and $A \in \mathfrak{A}$, then* $\mathrm{rp}(A) \sim \mathrm{rp}(A^*)$.

Proof. If $A = U(A^*A)^{1/2}$ is the polar decomposition described at the end of section 2, then $U \in \mathfrak{A}$ implements the equivalence $\mathrm{rp}(A) \sim \mathrm{rp}((A^*A)^{1/2})$.

Now the equality $||(A^*A)^{1/2}x||^2 = ||Ax||^2$ shows that $(A^*A)^{1/2}$ and A have the same null space, so (by §0, 2) that $[\mathrm{ran}((A^*A)^{1/2})]$ = $[\mathrm{ran}(A^*)]$, and the desired equivalence follows.

LEMMA 7. *If E and F are projections in a von Neumann algebra $\mathfrak{A}$, and if $EAF \neq 0$ for some $A \in \mathfrak{A}$, then there are non-zero subprojections $E_1 \leq E$ and $F_1 \leq F$ in $\mathfrak{A}$ with $E_1 \sim F_1$.*

Proof. Take $E_1 = \mathrm{rp}(EAF)$ and $F_1 = \mathrm{rp}(FA^*E)$. Then $E_1, F_1 \in \mathfrak{A}$ by Theorem 1 and $E_1 \sim F_1$ by Lemma 6.

COROLLARY 4. *$E\mathfrak{A}F \neq 0$ if and only if E and F have non-zero equivalent subprojections.*

Proof. The forward implication is Lemma 7. Conversely, if $E_1 = U^*U$ and $F_1 = UU^*$ are non-zero, so is EU^*F.

For $\mathfrak{A}$ a von Neumann algebra and $A \in \mathfrak{A}$, the *central cover*, denoted by C_A, of A is defined to be the GLB of all central projections $G \in \mathfrak{A}$ such that $GA = A$.

Remarks. (1) The *center* of $\mathfrak{A}$, namely $\mathfrak{A} \cap \mathfrak{A}'$, is itself a von Neumann algebra, so C_A is central.

(2) Each central projection $G \in \mathfrak{A}$ with $GA = A$ satisfies $\mathrm{ran}(G) \supset \mathrm{ran}(A)$, so $\mathrm{ran}(C_A) \supset \mathrm{ran}(A)$ and $C_A \cdot A = A$ (C_A is the smallest central projection containing $\mathrm{rp}(A)$).

(3) $\mathrm{ran}(C_A) = [\mathfrak{A}\mathfrak{A}' \, \mathrm{ran}(A)]$ (*Proof.* Each central projection containing $\mathrm{rp}(A)$ is invariant under $\mathfrak{A}$ and $\mathfrak{A}'$, and so $\mathrm{ran}(C_A) \supset [\mathfrak{A}\mathfrak{A}' \, \mathrm{ran}(A)]$. But the latter is clearly invariant under

$\mathcal{A}$ and $\mathcal{A}'$, so the projection onto it is central (by §0, 7), hence equals C_A by (2).

LEMMA 8. *Projections E and F in a von Neumann algebra $\mathcal{A}$ have non-zero equivalent subprojections if and only if $C_E \cdot C_F \neq 0$.*

Proof. If $C_E \cdot C_F = 0$, then $EAF = EC_E AC_F F = 0$, so no subprojection of E is equivalent to a subprojection of F by Corollary 4.

Conversely, if E and F have no non-zero equivalent subprojections, then $E\mathcal{A}F = 0$ by Corollary 4. Thus $E\mathcal{A}\mathcal{A}'F = 0$, so that $E \cdot C_F = 0$ by remark (3) above, and hence $C_E \cdot C_F = 0$.

Exercise. If $E \sim F$ and if G is a central projection, then $EG \sim FG$.

THE COMPARISON THEOREM. *For any two projections E and F in a von Neumann algebra $\mathcal{A}$, there is a central projection $G \in \mathcal{A}$ such that $EG \leqslant FG$ and $EG^\perp \geqslant FG^\perp$.*

Proof. Let $\{E_n\}$ and $\{F_n\}$ be a maximal pair of families of orthogonal projections such that $E_n \leqslant E$, $F_n \leqslant F$ and $E_n \sim F_n$. Set $E_1 = \text{LUB } E_n$ and $F_1 = \text{LUB } F_n$. By Additivity of Equivalence, $E_1 \sim F_1$. Let $E_2 = E - E_1$ and $F_2 = F - F_1$. If E_2 and F_2 had non-zero equivalent subprojections, we could add these to enlarge $\{E_n\}$ and $\{F_n\}$, thus contradicting maximality. Hence, by Lemma 8, $C_{E_2} \perp C_{F_2}$. Take $G = C_{F_2}$, and let $E_1 = U^*U$, $F_1 = UU^*$ with $U \in \mathcal{A}$. Then $(UG)^*(UG) = E_1 G$ and $(UG)(UG)^* = F_1 G$ so
$$EG = E_1 G + E_2 G = E_1 G + E_2 C_{E_2} C_{F_2} = E_1 G \sim F_1 G \leqslant FG \text{ and}$$
$$FG^\perp = F_1 G^\perp + F_2 G^\perp = F_1 G^\perp + F_2 (I - C_{F_2}) = F_1 G^\perp \sim E_1 G^\perp \leqslant EG^\perp.$$

Exercise. Show that for two projections E and F in a von Neumann algebra $\mathcal{A}$, there is a central projection $G \in \mathcal{A}$ with $EG \leqslant FG$ and $E^\perp G^\perp \leqslant F^\perp G^\perp$ (*Hint.* Compare $E \wedge F^\perp$ and $E^\perp \wedge F$ and use Corollary 3(2).)

§4. Classifications of von Neumann Algebras

A projection E in a von Neumann algebra $\mathcal{A}$ is said to be *finite* if whenever $E \sim F \leq E$ for a projection $F \in \mathcal{A}$, it follows that $F = E$. Thus a finite projection is one which is not equivalent to a proper part of itself. On the other hand, a projection which is equivalent to one of its proper subprojections is said to be *infinite*. We call the algebra $\mathcal{A}$ itself *finite* if the identity projection $I \in \mathcal{A}$ is finite.

Exercises. (1) If E is finite and $F \leq E$, then F is finite.

(2) If $\mathcal{A}$ is abelian, then $\mathcal{A}$ is finite (*Hint.* Note that $\sim$ reduces to $=$ in this case).

(3) If $\mathcal{A}$ is the algebra of all $n \times n$ complex matrices (n finite), then $\mathcal{A}$ is finite.

(4) The LUB of any family of central projections is central (*Hint.* Use remark (1) preceding Lemma 8 together with the proof of Proposition 2).

LEMMA 9. *Any family of finite projections whose central covers are pairwise orthogonal has finite LUB. Furthermore, the LUB of finite central projections is finite (and central).*

Proof. Let $\{E_n\}$ be a family of finite projections with $C_{E_n} \perp C_{E_m}$ for $n \neq m$ and put $E = \text{LUB } E_n$. Suppose $E \sim F \leq E$. Since $E - E_n < \underset{m \neq n}{\text{LUB }} C_{E_m}$, we have $EC_{E_n} - E_n = (E - E_n)C_{E_n} = 0$.

Further, $E_n - FC_{E_n} = (E - F)C_{E_n} \geq 0$ so that $E_n = EC_{E_n} \sim FC_{E_n} \leq E_n$ by the exercise after Lemma 8. From finiteness of E, we obtain $E_n = FC_{E_n} \leq F$ so $E \leq F$ and $F = E$.

Finally if $\{E_n\}$ is a family of finite central projections with LUB E and if $E \sim F \leq E$, then $E_n = EE_n \sim FE_n$, again by the exercise just cited, so $E_n = FE_n \leq F$ and $E \leq F$. Thus $F = E$.

Exercise. The second part of Lemma 9 follows from the first (*Hint*. Consider a maximal orthogonal family of central subprojections of the E_n's.)

A projection $E \in \mathcal{C}$ is *abelian* if $E\mathcal{C}E$ is commutative. A von Neumann algebra $\mathcal{C}$ is *type I* if every non-zero central projection contains a non-zero abelian projection. We say that $\mathcal{C}$ is *type II* if it has no abelian projections except zero and every non-zero central projection contains a non-zero finite projection. Call $\mathcal{C}$ *type III* if $\mathcal{C}$ contains no non-zero finite projection. A von Neumann algebra which is both finite and of type II is said to be of *type II$_1$*.

Exercises. (1) Every abelian projection is finite.

(2) If E is abelian and $F \leq E$, then F is abelian (*Hint*. $F\mathcal{C}F$ is isomorphic with a subalgebra of $E\mathcal{C}E$).

(3) Every abelian von Neumann algebra is type I.

(4) The ring $\mathcal{B}(\mathcal{H})$ of all bounded operators on $\mathcal{H}$ is type I. It is finite if and only if dim $\mathcal{H} < \infty$.

Several other subdivisions and terms in current use are listed below.

(1) Type $I_n \equiv$ (type I and finite).

(2) Discrete $\equiv$ type I.

(3) Properly infinite $\equiv$ (no finite central projection $\neq 0$).

(4) Semifinite $\equiv$ (every non-zero central projection contains a non-zero finite projection).

(5) Type $I_\infty \equiv$ (type I and properly infinite).

(6) Type $II_\infty \equiv$ (type II and properly infinite).

(7) Continuous $\equiv$ (no abelian projection $\neq 0$).

(8) Purely infinite $\equiv$ type III.

Each central projection E in a von Neumann algebra $\mathcal{A}$ gives rise to a direct sum decomposition $\mathcal{A} = \mathcal{A}E \oplus \mathcal{A}E^\perp$. The direct summands are ideals as well as von Neumann algebras.

Exercises. (1) If $\mathcal{A} = \mathcal{B} + \mathcal{C}$ (vector space sum), where $\mathcal{B}$ and $\mathcal{C}$ are von Neumann algebras contained in $\mathcal{A}$ with $\mathcal{BC} = 0$, then there is a unique central projection $E \in \mathcal{A}$ with $\mathcal{B} = \mathcal{A}E$ and $\mathcal{C} = \mathcal{A}E^\perp$ (*Hint.* Take E to be the LUB of all projections in $\mathcal{B}$.)

(2) If G is a central projection in $\mathcal{A}$ and H is a projection in $\mathcal{A}$ with $H \leqslant G$, then H is central in $\mathcal{A}$ if and only if H is central in $\mathcal{A}G$.

A von Neumann algebra whose center reduces to scalar multiples of the identity is called a *factor*.

THEOREM 3. *Any von Neumann algebra decomposes uniquely into five direct summands as follows:*

(1) *Type* I_n.

(2) *Type* I_∞.

(3) *Type* II_1.

(4) *Type* II_∞.

(5) *Type III*.

A factor has one and only one of these five types.

Proof. For a von Neumann algebra $\mathcal{A}$, let E be the LUB of all central projections P such that $\mathcal{A}P$ is type I, and let F be the LUB of all central projections Q for which $\mathcal{A}Q$ is type II. Then $EF = 0$ and if $G = I - E - F$, G is type III (the central cover of a finite projection in $\mathcal{A}G$ would be $\leqslant G$ and also $\leqslant E + F$, since the same projections annihilate a projection and its central cover). If H is the LUB of all finite central projections, the last half of Lemma 9, the definitions and exercise (2) above show that the required summands are $\mathfrak{A}EH$ (type I_n), $\mathfrak{A}EH^\perp$ (type I_∞), $\mathfrak{A}FH$ (type II_1), $\mathfrak{A}FH^\perp$ (type II_∞) and $\mathfrak{A}G$ (type III). Since a factor has only one non-zero central projection, the last statement is clear.

A projection E in a von Neumann algebra $\mathfrak{A}$ is said to be *faithful* if $C_E = I$.

Exercises. (1) A projection E is faithful if and only if no non-zero central projection of $\mathfrak{A}$ annihilates E.

(2) If $E =$ LUB E_n, then $C_E =$ LUB C_{E_n} (*Hint.* Use remark (2) before Lemma 8).

PROPOSITION 3. *A von Neumann algebra $\mathfrak{A}$ is type I if and only if $\mathfrak{A}$ contains a faithful abelian projection.*

Proof. With $\mathfrak{A}$ type I, choose a family of non-zero abelian projections $\{E_n\}$ maximal with respect to the property $C_{E_n} \perp C_{E_m}$, for $n \neq m$, and put $E =$ LUB E_n. By maximality and the type I hypothesis, E is faithful. Putting $G_n = C_{E_n}$, we have LUB $G_n = I$, by the second exercise before the proposition.

The mapping $A \mapsto (AG_n)$ is a $*$-algebra homomorphism of $\mathfrak{A}$ into the (algebraic) direct product $\Pi \mathfrak{A} G_n$. We assert that this map is injective. For if $AG_n = 0$ for all n, and if x is any vector in $\mathcal{H}$, then x is the limit of vectors in the (unclosed) linear span of $\cup$ ran(G_n), so $Ax = 0$ and $A = 0$, proving the assertion.

Now the image of $E\mathfrak{A}E$ under this mapping is contained in the product of the $E_n \mathfrak{A} E_n$'s (since $EG_n = E_n$ and $(EAE)G_n = E_n A E_n$). Since the latter are abelian, $E\mathfrak{A}E$ is too. Thus E is a faithful abelian projection.

For the converse, let E be a faithful abelian projection, and let

$G \neq 0$ be a central projection in $\mathcal{A}$. We must show that the direct summand $\mathcal{A}G$ contains a non-zero abelian projection.

But $EG \neq 0$ since E is faithful (exercise (1) before the proposition), so $C_E \cdot G \neq 0$. By Lemma 8, E and G have non-zero equivalent subprojections, say $0 \neq E_1 \leqslant E$ and $0 \neq G_1 \leqslant G$ with $E_1 \sim G_1$. But E_1 is abelian, and hence G_1 is too (see exercise (2) after the definition of types).

Exercise. A faithful abelian projection is *maximal abelian* (i.e., is contained in no strictly larger abelian projection).

PROPOSITION 4. *A von Neumann algebra $\mathcal{A}$ is semifinite (i.e., has no type III summand) if and only if $\mathcal{A}$ contains a faithful finite projection.*

Proof. Copy the proof of Proposition 3, replacing 'abelian' by 'finite', and using Lemma 9 in place of the direct product argument.

§5. The Commutant and Type I

The interplay between a von Neumann algebra and its commutant leads to several of the more profound aspects of the subject. In this section we pursue a fragment of commutant theory leading to a revealing and useful description of type I algebras.

LEMMA 10. *Let $\mathcal{A}$ be a von Neumann algebra acting on the Hilbert space $\mathcal{H}$ with $I \in \mathcal{A}$, and let $E \in \mathcal{A}$ be a projection. Then*

$$\mathrm{ran}(C_E) = [\mathcal{A}\,\mathrm{ran}(E)].$$

Proof. By remark (3) before Lemma 8, $\mathrm{ran}(C_E) = [\mathcal{A}\mathcal{A}'\,\mathrm{ran}(E)]$. Now $\mathrm{ran}(E) \subset \mathcal{A}'\,\mathrm{ran}(E)$, since $I \in \mathcal{A}'$. Also if $x \in \mathcal{H}$ and $A' \in \mathcal{A}'$, then $A'Ex = EA'x$, so $\mathcal{A}'\,\mathrm{ran}(E) = \mathrm{ran}(E)$, and the result follows.

Remarks. (1) If $E \in \mathcal{A}$ is a projection, then $\mathcal{A}'E \subset (E\mathcal{A}E)'E$. (*Proof.* $E\mathcal{A}E \subset \mathcal{A}$, so $\mathcal{A}' \subset (E\mathcal{A}E)'$ and $\mathcal{A}'E \subset (E\mathcal{A}E)'E$).

(2) Let $E \in \mathcal{A}$ be a projection and let G be any central projection in $\mathcal{A}$ with $G \geq E$ (e.g., $G = C_E$). Then $(\mathcal{A}G)'E = \mathcal{A}'E$ (*Proof.* Let $S \in (\mathcal{A}G)'E$. Then $S = DE$ with $D \in (\mathcal{A}G)'$. For $A \in \mathcal{A}$, $(DG)A = D(AG) = (AG)D = A(DG)$, so $DG \in \mathcal{A}'$. Thus $S = (DG)E \in \mathcal{A}'E$ and $(\mathcal{A}G)'E \subset \mathcal{A}'E$. But clearly $\mathcal{A}'E \subset (\mathcal{A}G)'E$ and equality holds).

THEOREM 4. *Let $\mathcal{A}$ be a von Neumann algebra acting on the Hilbert space $\mathcal{H}$ with $I \in \mathcal{A}$, and let $E \in \mathcal{A}$ be any projection. Then*

$$(E\mathcal{A}E)'E = \mathcal{A}'E.$$

Proof. Remark (1) before the theorem gives one inclusion, and for the other inclusion, it will be enough to show that $\mathcal{A}'E$ contains each unitary operator in $(E\mathcal{A}E)'E$ (in a von Neumann algebra, every operator is a linear combination of four unitaries in the algebra—see the second remark after Proposition 2).

Let U be unitary in $(E\mathcal{A}E)'E$. Then $U^*U = E = UU^*$ and $U = DE = ED$, where $D \in (E\mathcal{A}E)'$. For $B_1, \ldots, B_n \in \mathcal{A}$ and $x_1, \ldots, x_n \in \operatorname{ran}(E)$,

$$\left\|\sum_{i=1}^n B_i D x_i\right\|^2 = \left\|\sum_{i=1}^n B_i E D x_i\right\|^2 = \sum_{i,k=1}^n (B_i E D x_i | B_k E D x_k)$$
$$= \sum_{i,k=1}^n (D^*(EB_k^* B_i E) D x_i | x_k)$$
$$= \sum_{i,k=1}^n (D^* D (E B_k^* B_i E) x_i | x_k)$$
$$= \sum_{i,k=1}^n (U^* U (E B_k^* B_i E) x_i | x_k) = \left\|\sum_{i=1}^n B_i x_i\right\|^2.$$

Hence the linear map

$$\sum_{i=1}^n B_i x_i \mapsto \sum_{i=1}^n B_i D x_i$$

is well-defined and isometric on the (unclosed) span of $\mathcal{A}\operatorname{ran}(E)$. We define an operator D' on $\mathcal{H}$ as follows. Set

$$D'\left(\sum_{i=1}^n B_i x_i\right) = \sum_{i=1}^n B_i D x_i$$

(extended to $[\mathcal{A}\operatorname{ran}(E)]$ by continuity) and let $D' = 0$ on $[\mathcal{A}\operatorname{ran}(E)]^\perp$. Then for $A \in \mathcal{A}$,

$$AD'\left(\sum_{i=1}^{n} B_i x_i\right) = A\left(\sum_{i=1}^{n} B_i D x_i\right) = \sum_{i=1}^{n} AB_i D x_i$$
$$= D'\left(\sum_{i=1}^{n} AB_i x_i\right) = D'A\left(\sum_{i=1}^{n} B_i x_i\right),$$

so $AD' = D'A$ on $[\mathcal{Q} \operatorname{ran}(E)] = \operatorname{ran}(C_E)$ (Lemma 10). But clearly $D' \cdot C_E = C_E \cdot D' = D'$, so $AD' = D'A$ and $D' \in \mathcal{Q}'$.

Finally, if $x \in \mathcal{H}$, $D'Ex = DEx = Ux$, so $U = D'E \in \mathcal{Q}'E$ and the proof is complete.

Examples. (1) Let $\mathcal{Q} = \mathcal{B}(\mathcal{H})$ and let E be any non-zero projection. Then $(E\mathcal{Q}E)'E = \mathcal{Q}'E = \{aE\}$ (since $\mathcal{Q}' = \{aI\}$). This situation may be viewed matrically as follows.

$$E = \left(\begin{array}{cc|c} I & & 0 \\ & & 0 \\ \hline 0 & 0 & 0 \end{array}\right), \quad E\mathcal{Q}E = \left\{\left(\begin{array}{cc|c} & & 0 \\ & * & 0 \\ \hline 0 & 0 & 0 \end{array}\right)\right\}$$

and

$$(E\mathcal{Q}E)' = \left\{\left(\begin{array}{cc|c} & & 0 \\ aI & & 0 \\ \hline 0 & 0 & * \end{array}\right)\right\}.$$

(2) Let $\mathcal{Q}$ be a von Neumann algebra acting on $\mathcal{H}$ and let $\mathcal{M} = M_2(\mathcal{Q})$ be the algebra of all 2×2 matrices with entries from $\mathcal{Q}$. If

$$E = \begin{pmatrix} I & 0 \\ 0 & 0 \end{pmatrix},$$

then

$$E\mathcal{M}E = \left\{\begin{pmatrix} A & 0 \\ 0 & 0 \end{pmatrix}: A \in \mathcal{Q}\right\}$$

and

$$(E\mathcal{M}E)' = \left\{\begin{pmatrix} A' & 0 \\ 0 & B \end{pmatrix}: A' \in \mathcal{Q}', \ B \in \mathcal{B}(\mathcal{H})\right\},$$

also

$$\mathfrak{M}' = \left\{ \begin{pmatrix} A' & 0 \\ 0 & A' \end{pmatrix} : A' \in \mathfrak{A}' \right\}$$

(note that any matrix with entries from $\mathcal{B}(\mathcal{H})$ commuting with $\mathfrak{M}$ must commute, in particular, with the partial isometries

$$\begin{pmatrix} 0 & I \\ 0 & 0 \end{pmatrix}, \begin{pmatrix} 0 & 0 \\ I & 0 \end{pmatrix} \in \mathfrak{M}).$$

Hence

$$(E\mathfrak{M}E)'E = \left\{ \begin{pmatrix} A' & 0 \\ 0 & 0 \end{pmatrix} : A' \in \mathfrak{A}' \right\} = \mathfrak{M}'E.$$

For any set $\mathcal{S} \subset \mathcal{B}(\mathcal{H})$ and any projection $E \in \mathcal{B}(\mathcal{H})$, let

$$\mathcal{S}_E = \{ES|_{\mathrm{ran}(E)} : S \in \mathcal{S}\} \subset \mathcal{B}(\mathrm{ran}(E)).$$

We call $\mathcal{S}_E$ the *compression*, or *induction*, of $\mathcal{S}$ to $\mathrm{ran}(E)$.

Exercises. Let $\mathfrak{A} \subset \mathcal{B}(\mathcal{H})$ be a $*$-algebra of operators, and let E be a projection in $\mathfrak{A}$ or $\mathfrak{A}'$. Then (1) $E\mathfrak{A}E = \{A \in \mathcal{B}(\mathcal{H}): A = EAE\}$, and (2) $E\mathfrak{A}E \cong \mathfrak{A}_E$ ($*$-algebra isomorphism) via the map $A \mapsto A|_{\mathrm{ran}(E}$

COROLLARY 5. *Let $\mathfrak{A}$ be a von Neumann algebra acting on the Hilbert space $\mathcal{H}$ with $I \in \mathfrak{A}$, and let $E \in \mathfrak{A}$ be any projection. Then*

$$(\mathfrak{A}_E)' = (\mathfrak{A}')_E.$$

(*Note.* The first commutant is taken in $\mathcal{B}(\mathrm{ran}(E))$, the second in $\mathcal{B}(\mathcal{H})$.)

Proof. Let $S \in (\mathfrak{A}_E)'$. Extend S to an operator T on $\mathcal{H}$ as follows:

$$Tx = \begin{cases} Sx & \text{if } x \in \mathrm{ran}(E) \\ 0 & \text{if } x \in \mathrm{ran}(E)^\perp \end{cases}$$

Then $T = TE = ET$. If $A \in \mathcal{A}$ and $x \in \mathcal{H}$, $T(EAE)x = S(EAE)x = S(EA|_{\text{ran}(E)})Ex = (EA|_{\text{ran}(E)})SEx = (EA|_{\text{ran}(E)})TEx = (EAE)Tx$. Thus $T \in (E\mathcal{A}E)'$. By Theorem 4, $T = A'E$, for some $A' \in \mathcal{A}'$. But then $S = T|_{\text{ran}(E)} = A'|_{\text{ran}(E)} \in (\mathcal{A}')_E$. The reverse inclusion is clear from the fact that $\mathcal{A}_E$ and $(\mathcal{A}')_E$ are commuting algebras.

Exercises. Without using Theorem 4, give a direct proof of the following statements.

(1) For a von Neumann algebra $\mathcal{A}$ with $I \in \mathcal{A}$, and $E \in \mathcal{A}'$ a projection, $(\mathcal{A}E)'E = E\mathcal{A}'E$.

(2) With $\mathcal{A}$ and $E \in \mathcal{A}'$ as in (1), show that $(\mathcal{A}_E)' = (\mathcal{A}')_E$ (*Hint.* Use (1) and proceed as in the proof of Corollary 5.)

(3) The conclusions of Theorem 4 and Corollary 5 are equivalent and imply the conclusions of the two exercises above, which are likewise equivalent.

Problem. Show that if $\mathcal{A}$ is a von Neumann algebra with $I \in \mathcal{A}$ and $E \in \mathcal{A}$ is a projection, then $\mathcal{A}_E$ is strongly closed. (From this and the three exercises above, Theorem 4 and Corollary 5 follow easily.)

Let $\mathcal{S} \subset \mathcal{B}(\mathcal{H})$ be any set of operators and let $\mathcal{X} \subset \mathcal{H}$ be any set of vectors. For notational convenience we define $\mathcal{S}\mathcal{X} = \{Sx : S \in \mathcal{S}, x \in \mathcal{X}\}$. We say that $\mathcal{X}$ is *cyclic* for $\mathcal{S}$ if $[\mathcal{S}\mathcal{X}] = \mathcal{H}$. We call $\mathcal{X}$ *separating* for $\mathcal{S}$ if $S \in \mathcal{S}$ and $Sx = 0$ for all $x \in \mathcal{X}$ imply $S = 0$. If $\mathcal{X} = \{x_0\}$ consists of one vector x_0, we apply these terms to the vector itself.

PROPOSITION 5. *Let $\mathcal{S}$ be a $*$-closed semigroup of operators on the Hilbert space $\mathcal{H}$ with I $\mathcal{H}$ (i.e., A^*, $AB \in \mathcal{S}$ if A, $B \in \mathcal{S}$). Then*

(1) $\mathcal{S}'$ *is a von Neumann algebra.*

(2) *A vector $x_0 \in \mathcal{H}$ is cyclic for $\mathcal{S}$ if and only if x_0 is separating for $\mathcal{S}'$.*

(3) For any vector $x_0 \in \mathcal{H}$, the projection E onto $[\mathcal{S}x_0]$ belongs to $\mathcal{S}'$.

(4) Let $\mathcal{A}$ be a von Neumann algebra on $\mathcal{H}$ with $I \in \mathcal{A}$, and let $x_0 \in \mathcal{H}$. Then x_0 is separating for $\mathcal{A}$ if and only if x_0 is cyclic for $\mathcal{A}'$.

(5) With $\mathcal{S}$ as above and $\mathcal{X} \subset \mathcal{H}$ any set of vectors, the projection E onto $[\mathcal{S}\mathcal{X}]$ is the smallest projection in $\mathcal{S}'$ whose range contains $\mathcal{X}$.

(6) Let $\mathcal{S}$ be as above and let $E \in \mathcal{B}(\mathcal{H})$ be a projection. Then ran(E) is invariant under $\mathcal{S}$ if and only if $E \in \mathcal{S}'$.

Proof. The first statement is clear from the opening remarks of §2. To prove (2), let x_0 be cyclic and suppose that $S' \in \mathcal{S}'$ with $S'x_0 = 0$. Then $S'Sx_0 = SS'x_0 = 0$, for each $S \in \mathcal{S}$, and hence S' vanishes on $[\mathcal{S}x_0] = \mathcal{H}$, so $S' = 0$. Conversely, if E is the projection onto $[\mathcal{S}x_0]$, then $Ex_0 = x_0$, since $I \in \mathcal{S}$. Now if $S, T \in \mathcal{S}$, then $STx_0 \in \mathcal{S}x$ by the semigroup property, so that $\mathcal{S}x_0$ is invariant under $\mathcal{S}$, as is, therefore $[\mathcal{S}x_0]$. This means, however (§0, 6), that $ESE = SE$, for each $S \in \mathcal{S}$, so that also $ES^*E = S^*E$, because $\mathcal{S}$ is *-closed. Thus $ES = ESE = SE$ on taking adjoints, and $E \in \mathcal{S}'$. Now $(I-E)x_0 = x_0 - x_0 = 0$, and because $I - E \in \mathcal{S}$ and x_0 is separating for $\mathcal{S}'$, $E = I$, or $[\mathcal{S}x_0] = \mathcal{H}$, so x_0 is cyclic for $\mathcal{S}$. This proves (2), and, incidentally, (3).

To prove (4), let E be the projection onto $[\mathcal{A}'x_0]$. Then by (3), $E \in \mathcal{A}'' = \mathcal{A}$ (Double Commutant Theorem) $Ex_0 = x_0$ (since $I \in \mathcal{A}'$), and $(I-E)x_0 = x_0 - x_0 = 0$, so $E = I$ if x_0 is separating for $\mathcal{A}$. Thus x_0 is cyclic for $\mathcal{A}'$. The other implication in (4) follows from (2) and the fact that $\mathcal{A}'' = \mathcal{A}$.

To see that (5) holds, let $F \in \mathcal{S}'$ be a projection with ran(F) $\supset \mathcal{X}$. If $S \in \mathcal{S}$ and $x \in \mathcal{X}$, then $Sx = SFx = F(Sx)$, so $\mathcal{S}\mathcal{X} \subset$ ran(F) and therefore $[\mathcal{S}\mathcal{X}] \subset$ ran(F) or $E \leq F$.

The proof of (6) follows from that of (2).

Remark. Parts (2), (3) and (4) are easily seen to hold if the single vector x_0 is replaced by an arbitrary subset $\mathfrak{X} \subset \mathcal{H}$.

Let $\mathfrak{A}$ be a von Neumann algebra acting on the Hilbert space $\mathcal{H}$. We say that $0 \neq x_0 \in \mathcal{H}$ is a *trace vector* for $\mathfrak{A}$ if $(ABx_0|x_0) = (BAx_0|x_0)$, for all operators $A, B \in \mathfrak{A}$. Notice that any vector $x_0 \neq 0$ is a trace vector if $\mathfrak{A}$ is abelian.

Exercise. If x_0 is a cyclic trace vector for $\mathfrak{A}$, then x_0 is separating for $\mathfrak{A}$.

Now assume that $\mathfrak{A}$ is a von Neumann algebra acting on $\mathcal{H}$ with cyclic trace vector $x_0 \in \mathcal{H}$. For $A \in \mathfrak{A}$ define a mapping

$$J: Ax_0 \mapsto A^*x_0.$$

Since x_0 is also separating (by the exercise above), J is well-defined. For $A, B \in \mathfrak{A}$ put $a = Ax_0$ and $b = Bx_0$. Then $(Ja|Jb) = (A^*x_0|B^*x_0) = (BA^*x_0|x_0) = (A^*Bx_0|x_0) = (Bx_0|Ax_0) = (b|a)$. Also $J^2a = J(A^*x_0) = Ax_0 = a$. Since $||Ja|| = ||a||$, J extends by continuity to a *conjugation* J on $\mathcal{H}$, i.e.,

(i) $J^2 = I$.
(ii) $(Jx|Jy) = (y\ x)$, for all $x, y \in \mathcal{H}$.

Exercise. For any conjugation, the following hold:

(iii) $(Jx|y) = (Jy|x)$.
(iv) $J(x + y) = Jx + Jy$.
(v) $J(\lambda x) = \bar{\lambda}(Jx)$.

(for all vectors $x, y \in \mathcal{H}$ and λ complex).

LEMMA 11. *With the assumptions and notation above,*

$$J(Tx_0) = T^*x_0, \qquad \text{if } T \in \mathfrak{A}'.$$

Proof. Take any $A \in \mathfrak{A}$. Then $Tx_0 = \lim A_n x_0$, for a suitable sequence of operators $A_n \in \mathfrak{A}$, by cyclicity of x_0. Thus

$$(J(Tx_0)|Ax_0) = \lim(J(A_n x_0)|Ax_0) = \lim(A_n^* x_0|Ax_0)$$
$$= \lim(x_0|A_n A x_0) = \lim(x_0|AA_n x_0) = (A^* x_0|Tx_0)$$
$$= (T^*A^* x_0|x_0) = (A^* T^* x_0|x_0) = (T^* x_0|Ax_0).$$

Since $[\mathfrak{A}x_0] = \mathcal{H}$, $J(Tx_0) = T^* x_0$.

LEMMA 12. *With the assumptions and notation above,* $J\mathfrak{A}J = \mathfrak{A}'$.

Proof. Let $A, B, T \in \mathfrak{A}$. Then $(JAJ)BTx_0 = J(AT^*B^* x_0) = BTA^* x_0$. Also $B(JAJ)Tx_0 = BJAT^* x_0 = BTA^* x$, so that $JAJ \in \mathfrak{A}'$ by cyclicity of x_0. Now take $A', B', T' \in \mathfrak{A}'$. Using Lemma 11 and cyclicity of x_0, this time for $\mathfrak{A}'$ (see the exercise after Proposition 5 and part (4) of the latter), a similar computation shows, via the Double Commutant Theorem, that $JA'J \in \mathfrak{A}$.

COROLLARY 6. *If a von Neumann algebra* $\mathfrak{A}$ *has a cyclic trace vector, then* $\mathfrak{A}$ *is* $*$-*antiisomorphic to* $\mathfrak{A}'$.

Proof. Let J be the conjugation constructed above. Then $A \mapsto JA^*J$ is an algebra antiisomorphism of $\mathfrak{A}$ with $\mathfrak{A}'$ by Lemma 12. To see that this mapping preserves adjoints, note that $(JA^*Jx|y) = (JA^*Jx|J^2 y) = (Jy|A^*Jx) = (AJy|Jx) = (J^2 AJy|Jx) = (x|JAJy)$.

THEOREM 5. *If an abelian von Neumann algebra* $\mathfrak{A}$ *has a cyclic vector, then* $\mathfrak{A}$ *is maximal abelian, i.e.,* $\mathfrak{A} = \mathfrak{A}'$.

Proof. Let x_0 be a cyclic vector for $\mathfrak{A}$. Then x_0 is also a trace vector, and hence is separating. But $\mathfrak{A} \subset \mathfrak{A}' \cong \mathfrak{A}$ (antiisomorphic), by Corollary 6, so $\mathfrak{A}'$ is abelian and therefore $\mathfrak{A}' \subset \mathfrak{A}'' = \mathfrak{A}$.

THEOREM 6. *If a von Neumann algebra $\mathfrak{A}$ has a cyclic trace vector, then $\mathfrak{A}$ has finite type.*

Proof. The functional $\omega(A) = (Ax_0|x_0)$ is a *trace* on $\mathfrak{A}$, i.e.,

(i) ω is linear.

(ii) $\omega(A) \geq 0$ if $A \in \mathfrak{A}^+$.

(iii) $\omega(A^*A) = \omega(AA^*)$, for all $A \in \mathfrak{A}$.

(iv) $\omega(A) = 0$ and $A \in \mathfrak{A}^+$ imply $A = 0$ (to see this, observe that $\omega(A) = (Ax_0|x_0) = (A^{1/2}x_0|A^{1/2}x_0) = ||A^{1/2}x_0||^2$, so $\omega(A) = 0$ implies $A^{1/2}x_0 = 0$, which because x_0 is separating implies $A^{1/2} = 0$ and $A = 0$).

(v) $\omega(I) = 1$ (replace x_0 by $x_0/||x_0||$ if necessary).

Let $E \in \mathfrak{A}$ be a projection with $E \sim I$. Then $U^*U = I$ and $UU^* = E$ for some isometry $U \in \mathfrak{A}$. But $U^*U - UU^* = I - E \geq 0$ and by (iii), $\omega(I - E) = 0$. By (iv), $E = I$.

LEMMA 13. *Let $\mathfrak{A}$ be a von Neumann algebra acting on the Hilbert space $\mathcal{H}$ with $I \in \mathfrak{A}$, and let $E \in \mathfrak{A}$ be a faithful projection. Then $\mathfrak{A}' \cong \mathfrak{A}'E$.*

Proof. The map $A' \mapsto A'E$ ($A' \in \mathfrak{A}$) is clearly a surjective *-algebra homomorphism. By Lemma 10, $[\mathfrak{A} \operatorname{ran}(E)] = \mathcal{H}$. But if $A'E = 0$ and $x \in \mathcal{H}$, then $A'AEx = AA'Ex = 0$, so $A' = 0$, and hence the map is also injective.

THEOREM 7. *If $\mathfrak{A}$ is a type I von Neumann algebra, then $\mathfrak{A}'$ is also type I.*

Proof. First suppose $\mathfrak{A}$ is abelian and let $G \neq 0$ be a central projection. Take any vector $0 \neq x \in \operatorname{ran}(G)$ and let E be the projection onto $[\mathfrak{A}x]$. By Proposition 5(3), $E \in \mathfrak{A}'$. Also $0 \neq E \leq G$ and by Corollary 5 and the Double Commutant Theorem, $(\mathfrak{A}')_E = (\mathfrak{A}_E)'$. But $\mathfrak{A} \subset \mathfrak{A}'$, so $\mathfrak{A}_E \subset (\mathfrak{A}')_E = (\mathfrak{A}_E)'$ and since x is cyclic for $\mathfrak{A}_E$ on ran(E), Theorem 5 gives $\mathcal{A}_E = (\mathcal{A}_E)'$. Thus $E\mathcal{A}'E \cong (\mathcal{A}')_E = (\mathcal{A}_E)' = \mathcal{A}_E$ by the second exercise before Corollary 5, so E is an abelian projection for $\mathcal{A}'$.

For the general case, let F be a faithful abelian projection in $\mathcal{A}$ (Proposition 3). Then $\mathcal{A}' \cong \mathcal{A}'F \cong (\mathcal{A}')_F = (\mathcal{A}_F)'$ by Lemma 13, the second exercise before Corollary 5, and by Corollary 5. Also $\mathcal{A}_F \cong F\mathcal{A}F$ (abelian), by the exercise just cited. By the abelian case, $\mathcal{A}'$ is type I.

THEOREM 8. *A von Neumann algebra is type I if and only if it is *-isomorphic to a von Neumann algebra with abelian commutant.*

Proof. Let $\mathcal{A}$ be a type I algebra. By Theorem 7, $\mathcal{A}'$ is also type I, hence by Proposition 3, $\mathcal{A}'$ contains a faithful abelian projection F. Then $(\mathcal{A}_F)' = (\mathcal{A}')_F \cong F\mathcal{A}'F$, as in the proof of Theorem 7, and the latter is abelian. By Lemma 13, $\mathcal{A} \cong \mathcal{A}_F$.

The other implication follows from the abelian case considered in Theorem 7.

PROPOSITION 6. *Let $\mathcal{A}$ be an abelian von Neumann algebra. Then $\mathcal{A}$ is *-isomorphic to a maximal abelian von Neumann algebra acting on a subspace of the space on which $\mathcal{A}$ acts.*

Proof. Since $\mathcal{A}'$ is type I (Theorem 7), we can find a faithful abelian projection $E \in \mathcal{A}'$ (Proposition 3). But $(\mathcal{A}')_E \cong E\mathcal{A}'E$ (exercise (2) before Corollary 5), and the latter is abelian. Hence $(\mathcal{A}')_E \subset ((\mathcal{A}')_E)' = \mathcal{A}_E$ by Corollary 5 and the Double Commutant Theorem. Now $\mathcal{A}_E$ is abelian, since $\mathcal{A}$ is and $E \in \mathcal{A}'$, and therefore $\mathcal{A}_E \subset (\mathcal{A}_E)'$, so $\mathcal{A}_E = (\mathcal{A}_E)'$ showing that $\mathcal{A}_E$ is maximal abelian. Further, $\mathcal{A} \cong \mathcal{A}E$ by Lemma 13, and $\mathcal{A}E \cong \mathcal{A}_E$ by the second exercise before Corollary 5.

§6. Type I Structure Theory

The structure theory of type I von Neumann algebras most nearly parallels the classical structure theorem of Wedderburn for finite dimensional semisimple algebras. The latter decompose into a finite direct sum of simple matrix algebras. In the case of a type I von Neumann algebra, one needs to extend the notion of direct sum to allow for an infinite number of summands, and to permit the use of matrices of infinite degree (see Appendix A). In general, the summands will no longer be simple algebras, nor in fact will their centers always be identifiable with the field of complex scalars. The attendant increase in complexity eventually leads to the need for direct integral theory.

We begin our structural analysis with a fundamental and highly useful relation between a von Neumann algebra and its center.

PROPOSITION 7. *Let $\mathcal{C}$ be a von Neumann algebra and let $\mathcal{Z} = \mathcal{C} \cap \mathcal{C}'$ be its center. Then for any projection $E \in \mathcal{C}$, the center of $E\mathcal{C}E$ is $\mathcal{Z}E$.*

Proof. First recall that for projections $E, F \in \mathcal{C}$ we have $E\mathcal{C}F = 0$ if and only if $C_E \cdot C_F = 0$, by Corollary 4 and Lemma 8.

Clearly $\mathcal{Z}E \subset \text{center}(E\mathcal{C}E)$. If F is a central projection in $E\mathcal{C}E$, then for $A \in \mathcal{C}$, $FA(E - F) = F(EAE)(E - F) = (EAE)F(E - F) = 0$, since $F < E$, so by the opening remark, $C_F \cdot C_{E-F} = 0$, and, in particular, $C_F \cdot (E - F) = 0$. Thus $F = C_F \cdot E \in \mathcal{Z}E$.

Since the centers of $E\mathcal{A}E$ and $\mathcal{Z}E$ are both von Neumann algebras containing the same projections, they coincide by Theorem 1.

COROLLARY 7. *Each central projection $F \in E\mathcal{A}E$ has the form $F = C_F \cdot E$.*

COROLLARY 8. *If E is abelian, $E\mathcal{A}E = \mathcal{Z}E$.*

COROLLARY 9. *If $\mathcal{A}$ is a factor and E is any non-zero projection in $\mathcal{A}$, then $E\mathcal{A}E$ is also a factor.*

COROLLARY 10. *If $\mathcal{A}$ is a factor, every abelian projection is minimal. If a von Neumann algebra has a faithful minimal projection, it is a type I factor.*

Proof. The first statement is clear from Corollary 8. If E is a faithful minimal projection in $\mathcal{A}$, then $E\mathcal{A}E$ is one dimensional and for any central projection $G \in \mathcal{A}$ we have $GE = E$ or $GE = 0$. If $GE = E$, $C_E \leq G$ and $G = I$. If $GE = 0$, $(I - G)E = E$ and $C_E \leq I - G$ so $G = 0$. Hence $\mathcal{A}$ is a factor (of type I).

The first statement of the next lemma is contained in the proof of Proposition 3. The second statement follows readily from the first (apply Zorn's Lemma, using the hint for the exercise after Lemma 9).

LEMMA 14. *Any family of abelian projections whose central covers are pairwise orthogonal has abelian LUB. Furthermore, the LUB of abelian central projections is abelian (and central).*

Remark. If $E \leq F$, then $C_E \leq C_F$. In particular, equivalent projections have the same central cover (*Proof.* If $E \leq F$, there is a partial isometry $U \in \mathcal{A}$ with $U^*U = E$ and $UU^* \leq F$. For $x \in \mathcal{H}$, put $y = Ux$. Then $AEx = AE^2x = AEU^*y = AU^*Fy$, so $\mathcal{A}\,\mathrm{ran}(E) \subset \mathcal{A}\,\mathrm{ran}(F)$. Now apply Lemma 10).

LEMMA 15. *In a type I von Neumann algebra, any projection contains an abelian projection with the same central cover. Hence $E\mathcal{C}E$ is type I if $\mathcal{C}$ is.*

Proof. Given a projection $0 \neq E \in \mathcal{C}$. By assumption there is an abelian projection $P \in \mathcal{C}$ with $C_P = I$. Since $E \neq 0$, $C_E \cdot C_P = C_E \neq 0$, so E and P have nonzero equivalent subprojections by Lemma 8. Thus we can find projections F and G with $0 \neq F \leqslant E$ and $0 \neq G \leqslant P$ so that $F \sim G$. But G is abelian and hence F is also. This shows that any nonzero projection contains a nonzero abelian projection.

Now let $\{E_n\}$ be a maximal family of abelian projections such that $E_n \leqslant E$ and the C_{E_n} are orthogonal. Set $H = \text{LUB } E_n$. By Lemma 14, H is abelian, and clearly $C_H \leqslant C_E$. If $C_H \neq C_E$, then $E(C_E - C_H) \neq 0$ would contain a nonzero abelian projection K. But then $C_K \cdot C_H = 0$ and $K \leqslant E$, contradicting maximality. Thus $C_H = C_E$.

Remark. An order preserving map on a complete lattice has a fixed point (*Proof.* Let φ be the map and let $p = \sup X$, where $X = \{g: g \leqslant \varphi(g)\}$. For $g \in X$, $g \leqslant p$ so that $g \leqslant \varphi(g) \leqslant \varphi(p)$ and $p \leqslant \varphi(p)$. Thus $\varphi(p) \leqslant \varphi(\varphi(p))$, so $\varphi(p) \in X$ and $\varphi(p) \leqslant p$).

'SCHRÖDER-BERNSTEIN THEOREM.' *Let E and F be projections in a von Neumann algebra. If $E \lesssim F$ and $F \lesssim E$, then $E \sim F$.*

Proof. Given $WW^* = E$, $W^*W \leqslant F$, $VV^* = F$ and $V^*V \leqslant E$. $G \leqslant F$, define

$$\varphi(G) = F - W^*(E - V^*GV)W.$$

It is easy to see that φ is order preserving on the complete lattice $\{G: G \text{ a projection} \leqslant F\}$, so there is a projection $P \leqslant F$ with $P = F - W^*(E - V^*PV)W$. The partial isometries PV and

$(E - V^*PV)W$ implement the equivalences $P \sim V^*PV$ and $F - P \sim E - V^*PV$, so by Additivity of Equivalence, $F \sim E$.

The above proof is due to Arnold Lebow [11].

LEMMA 16. *Two abelian projections in a von Neumann algebra which have the same central cover are equivalent.*

Proof. It is enough, by the preceding theorem, to show that if E and F are abelian projections and $C_E \leqslant C_F$, then $E \leqslant F$. By the Comparison Theorem, it is enough to consider the case where $F \leqslant E$, and hence the case where $F \leqslant E$. Then $F = C_F \cdot E$ by Corollary 7, since F is central in $E\mathcal{C}E$. Since $E \leqslant C_E \leqslant C_E$, we have $F = E$.

A von Neumann algebra is said to be *homogeneous* if it contains an orthogonal family $\{E_n\}$ of equivalent abelian projections with LUB $E_n = I$.

Exercises. (1) A homogeneous algebra is type I. (2) Every type I factor is homogeneous.

TYPE I STRUCTURE THEOREM. *Let $\mathcal{C}$ be a type I von Neumann algebra. Then there is a family $\{G_n\}$ of orthogonal central projections with LUB $G_n = I$ such that each $\mathcal{C}G_n$ is homogeneous.*

Proof. Let $\{G_n\}$ be a maximal family of orthogonal central projections such that $\mathcal{C}G_n$ is homogeneous. It suffices to show that $I - \text{LUB } G_n = 0$. By maximality, $\mathcal{C}(I - \text{LUB } G_n)$ contains no homogeneous direct summand. Hence it suffices to show that every type I algebra contains a non-trivial homogeneous summand.

With $\mathcal{C}$ type I, let $\{E_n\}$ be a maximal family of orthogonal faithful abelian projections in $\mathcal{C}$ (whose existence is guaranteed by Proposition 3) and set $E = \text{LUB } E_n$. Now $I - E$ cannot be faithful, for then it would contain a faithful abelian projection by Lemma 15, contrary to maximality. Set $H = I - C_{I-E}$. Then $(I - E)H = 0$ and

$H = HE \neq 0$. Since H commutes with each E_n, $H = HE = \text{LUB } HE_n$. By Lemma 16, the E_n's are equivalent, as are therefore the HE_n's, by the exercise after Lemma 8. Since HE_n is abelian, $\mathcal{A}H$ is a non-trivial homogeneous summand of $\mathcal{A}$, and the proof is complete.

Remarks. Given any family $\{\mathcal{A}_n\}$ of von Neumann algebras, there is a von Neumann algebra $\mathcal{A}$ called the *direct sum of the $\mathcal{A}_n$'s*, written $\mathcal{A} = \Sigma + \mathcal{A}_n$ (see Appendix A). If $\mathcal{A}_n$ acts on the Hilbert space $\mathcal{H}_n$, then $\mathcal{A}$ acts on the direct sum $\mathcal{H} = \Sigma \oplus \mathcal{H}_n$ and is characterized by the properties: (i) the $\mathcal{A}_n$'s are spatially isomorphic to pairwise orthogonal ideals in $\mathcal{A}$, and (ii) if $\mathcal{U}$ denotes the set-theoretic union of these ideals, then $\mathcal{A} = \mathcal{U}''$.

Thus, in the preceding structure theorem, the type I algebra $\mathcal{A}$ is spatially isomorphic to the direct sum $\Sigma \oplus \mathcal{A}G_n$ of the homogeneous algebras $\mathcal{A}G_n$.

The homogeneous summands are further analyzed as follows. For $\mathcal{A}$ homogeneous, there is a unique cardinal number n called the *degree of homogeneity*, this being the cardinality of any orthogonal decomposition of the identity operator $I \in \mathcal{A}$ into equivalent abelian projections. If E denotes one of the latter, then $\mathcal{A}$ is spatially isomorphic to the von Neumann algebra $M_n(E\mathcal{A}E)$ of all $n \times n$ matrices with entries from $E\mathcal{A}E$ which act as bounded operators on the Hilbert space direct sum of n copies of $\text{ran}(E)$ with itself. The algebra $E\mathcal{A}E$ turns out to be $*$-isomorphic with the center of $\mathcal{A}$. This, and related constructions, are described more fully in Appendix A.

§7. Some General Structure Theory

We now drop the type I hypothesis and concentrate on continuous algebras. We are interested in the question of when a von Neumann algebra $\mathcal{A}$ is $*$-isomorphic to $M_n(\mathcal{A})$, the algebra of all $n \times n$ matrices with entries from $\mathcal{A}$ which act as bounded operators ($2 \leqslant n \leqslant \aleph_0$). Knowledge of this situation will be invaluable in our later examination of the generator problem (§10).

We begin with a fundamental property of continuous algebras.

THE 'HALVING LEMMA'. *Let $\mathcal{A}$ be a von Neumann algebra which has no type I summand. Then any non-zero projection in $\mathcal{A}$ can be split into two orthogonal equivalent halves.*

Proof. Given a projection $0 \neq E \in \mathcal{A}$, let $(\{F_n\}, \{G_n\})$ be a maximal pair of families of orthogonal projections such that $F_n, G_n \leqslant E$, $F_n \perp G_n$ and $F_n \sim G_n$. Set $F = \mathrm{LUB}\, F_n$ and $G = \mathrm{LUB}\, G_n$. Then $F \perp G$, and by Additivity of Equivalence, $F = G$. Our task is to show that $E = F + G$, and for this it is enough to show that if a projection P is not abelian, then we can find subprojections $0 \neq P_1 \leqslant P$ and $0 \neq P_2 \leqslant P$ in $\mathcal{A}$ with $P_1 \perp P_2$ and $P_1 \sim P_2$. But if P is not abelian, there is a non-central projection $Q \in P\mathcal{A}P$. If $C_Q C_{P-Q} = 0$, then $Q = C_Q \cdot P$ is central in $P\mathcal{A}P$, a contradiction. Thus Q and $P - Q$ have non-zero equivalent subprojections by Lemma 8.

Finally, if $P = E - (F + G) \neq 0$, P is non-abelian since $\mathfrak{A}$ has no non-zero abelian projections. The above reasoning would then contradict maximality.

Exercise. An abelian projection cannot be 'halved.'

Remark. If E is a projection in a von Neumann algebra $\mathfrak{A}$, then two projections in $E\mathfrak{A}E$ are equivalent in $E\mathfrak{A}E$ if and only if they are equivalent in $\mathfrak{A}$ (*Proof.* Let F, $G \leqslant E$ be projections. If $U \in \mathfrak{A}$ satisfies $F = U^*U$ and $G = UU^*$, then $EUE = E(GUF)E = GUF = U$, so $U \in E\mathfrak{A}E$ and $F \approx G$ in $E\mathfrak{A}E$. The other implication is clear).

COROLLARY 11. *If $\mathfrak{A}$ is a continuous von Neumann algebra, and if $E \in \mathfrak{A}$ is a non-zero projection, then $E\mathfrak{A}E$ is also continuous.*

Proof. If $E\mathfrak{A}E$ contained a nonzero abelian projection F, then $F\mathfrak{A}F = F(E\mathfrak{A}E)F$, so that F would be abelian in $\mathfrak{A}$ as well, contrary to the assumption that $\mathfrak{A}$ is continuous.

LEMMA 17. *Let $\mathfrak{A}$ be a von Neumann algebra and let G be a central projection in $\mathfrak{A}$.*

(1) *If E is a finite projection in $\mathfrak{A}$, then EG is finite in $\mathfrak{A}G$.*

(2) *If E is a finite projection in $\mathfrak{A}G$, then E is finite in $\mathfrak{A}$.*

Proof. (1) Given a partial isometry $U \in \mathfrak{A}G$ with $EG = U^*U$ and $UU^* \leqslant EG$. We must show that $UU^* = EG$. Now $U(EG) = UU^*U = U$. Further, $EG - U^*(EG)U = U^*(I - EG)U \geqslant 0$ and $EG = U^*(UU^*)U \leqslant U^*(EG)U$ so that $U^*(I - EG)U = 0$. Putting $X = (I - EG)U$, we have $X^*X = 0$, so $X = 0$ and therefore $U = (EG)U$. Now put $V = U + E(I-G)$. Then $V^*V = EG + E(I - G) = E$ and since $V \in E\mathfrak{A}E$, the latter being finite, and $VV^* < E$, we have $VV^* = E$. But $VV^* = UU^* + E(I - G)$, so $E = UU^* + E - EG$ and $UU^* = EG$.

(2) Given a partial isometry $U \in \mathfrak{A}$ with $E = U^*U$ and $UU^* \leqslant E$.

SOME GENERAL STRUCTURE THEORY   47

Then $U = UU^*U = UE$, so $UG = UEG = UE = U$ and therefore $U \in \mathfrak{A}G$ and finiteness implies $UU^* = E$.

Exercise. Prove Lemma 17 with 'finite' replaced by 'abelian.'

COROLLARY 12. *Any non-zero direct summand of a properly infinite von Neumann algebra is properly infinite.*

Proof. Let $\mathfrak{A}$ be properly infinite and let $G \neq 0$ be a central projection in $\mathfrak{A}$. If E is a finite central projection in $\mathfrak{A}G$, then $E = C_n \cdot G$ by Corollary 7, so E is central in $\mathfrak{A}$. By Lemma 17(2), E is finite in $\mathfrak{A}$, so $E = 0$. Thus $\mathfrak{A}G$ is properly infinite.

THEOREM 9. *If $\mathfrak{A}$ is a properly infinite von Neumann algebra, then $\mathfrak{A}$ contains an infinite orthogonal sequence $\{E_n\}_{n=1}^\infty$ of equivalent projections with* LUB $E_n = I$.

Proof. Since $\mathfrak{A}$ is not finite, it contains a partial isometry X with $XX^* = I$ and $X^*X \neq I$. Write $G_n = (X^*)^n X^n$, $Y_n = (X^*)^n X^{n+1}$ and $E_n = G_n - G_{n+1}$. Then the E_n's are orthogonal non-zero projections in $\mathfrak{A}$ with $E_n \approx E_{n+1}$ via the partial isometry $Y_n - Y_{n+1}$.

Hence there is an infinite orthogonal family $\{E_n\}$ of equivalent non-zero projections, and we may suppose the family to be maximal. Put $E = $ LUB E_n and let $F = I - E$. By the Comparison Theorem, there is a central projection $H \in \mathfrak{A}$ such that $FH \leqslant E_1 H$ and $E_1(I - H) \leqslant F(I - H)$, where E_1 is any fixed projection in the family $\{E_n\}$. Now $EH \neq 0$, for otherwise we would have $E_1 = E_1(I - H) \leqslant F(I - H) \leqslant F = I - E$, so $E_1 \leqslant I - E$, contradicting maximality of the family $\{E_n\}$.

Next note that LUB $E_n H \sim $ LUB$_{n \neq 1} E_n H$, since LUB $E_n \sim $ LUB$_{n \neq 1} E_n$ by Additivity of Equivalence. Hence $H = EH + (I-E)H = $ (LUB $E_n H$)+ $FH \sim $ (LUB$_{n \neq 1} E_n H$) + $FH \leqslant $ (LUB$_{n \neq 1} E_n H$)+ $E_1 H = $ LUB $E_n H = EH$. Thus $H \leqslant EH$ and since $EH \leqslant H$, the

Schröder-Bernstein Theorem implies $H \sim EH$. This equivalence induces an orthogonal decomposition $H = \text{LUB } H_n$ with $H_n \sim E_n H$. Now divide the family $\{H_n\}$ into $\aleph_0$ disjoint subfamilies, each having the same cardinality as $\{H_n\}$. Adding the projections in each of these subfamilies yields an infinite (countable) sequence of equivalent projections with LUB H.

We have shown that any properly infinite algebra contains a non-zero summand having the desired property. Corollary 12, a Zorn's Lemma argument and a final piecing together (using Additivity of Equivalence) completes the proof.

COROLLARY 13. *Every properly infinite von Neumann algebra contains a projection E with $I \sim E \sim I - E$.*

Proof. By Theorem 9, we can write $I = \text{LUB } E_n$ where the E_n's are orthogonal equivalent projections. Set $E = \text{LUB } E_n$ ($n = 1, 3, 5, \ldots$). By Additivity of Equivalence, $I \sim E$ and $I \sim I - E$.

Remark. If $E \in \mathcal{A}$ is a projection with $E \sim I$, then $\mathcal{A} \cong E\mathcal{A}E$ (spatially). In fact, the mapping $A \mapsto VAV^*$, where $V^*V = I$ and $VV^* = E$, is a spatial *-isomorphism. In particular, if $\mathcal{A}$ is properly infinite, so is $E\mathcal{A}E$.

COROLLARY 14. *Let $\mathcal{A}$ be a properly infinite von Neumann algebra and let n be any cardinal number $\leq \aleph_0$. Then there is a spatial *-isomorphism*

$$\mathcal{A} \cong M_n(E\mathcal{A}E)$$

where $E \in \mathcal{A}$ is any projection with $I \sim E \sim I - E$. In particular,

$$\mathcal{A} \cong M_n(\mathcal{A}) \quad \text{(spatially)}.$$

Proof. If $I \sim E \sim I - E$, we can (by the remark above) write $I - E = E_2 + F_2$, $F_2 = E_3 + F_3$, etc., where E_n, $F_n \in \mathcal{A}$ are orthogonal projections, each $E_n \sim E$, LUB $E_n = I$ and the E_n's are pairwise orthogonal. Put $E_1 = E$ and let $U_n \in \mathfrak{A}$ be a partial isometry with $U_n^* U_n = E_1$ and $U_n U_n^* = E_n$. Let $V \in \mathfrak{A}$ be an isometry with $VV^* = E_1$. Then the last isomorphism is given by

$$A \to ((U_i V)^* A (U_j V)).$$

Problem. Characterize those finite type II algebras $\mathfrak{A}$ for which $\mathfrak{A} \cong M_2(\mathfrak{A})$ (*-isomorphism). This is known to hold for *hyperfinite* II_1-factors (see §11 and [1; p.272, Th.2 and p.281, Prop.3]).

Remark. If E, F and G are projections in a von Neumann algebra with $E \leqslant F \sim G$, then $E \lesssim G$ (*Proof.* Let $F = U^*U$ and $G = UU^*$. Put $V = UE$. Then $V^*V = EU^*UE = EFE = E$ and $VV^* = UEU^* \leqslant UFU^* = UU^*UU^* = G$.)

PROPOSITION 8. *Let $\mathfrak{A}$ be a von Neumann algebra, and let E, $F \in \mathfrak{A}$ be projections with $E \sim I - E$ and $F \sim I - F$. Then $E \sim F$.*

Proof. By the exercise after the Comparison Theorem, we can find a central projection $G \in \mathfrak{A}$ with $EG \lesssim FG$ and $(I - E)(I - G) \lesssim (I - F)(I - G)$. Then $E = EG + E(I - G) \lesssim FG + E(I - G) \sim FG + (I - E)(I - G) \approx (I - F)G + (I - E)(I - G) \lesssim (I - F)G + (I - F)(I - G) = I - F \sim F$, and by the above remark, $E \lesssim F$. By symmetry, $F \lesssim E$ and by the Schröder-Bernstein Theorem, $F \sim E$.

COROLLARY 15. *Let $\mathfrak{A}$ be a von Neumann algebra, and let E, $F \in \mathfrak{A}$ be projections with $E \sim I - E$ and $F \sim I - F$. Then there is a unitary operator $U \in \mathfrak{A}$ with $E = U^*FU$.*

Proof. By Proposition 8, $E \sim F$ and $I - E \sim I - F$. Let V, W be partial isometries with $E = V^*V$, $F = VV^*$, $I - E = W^*W$ and $I - F = WW^*$. Put $U = V + W$.

Since $FW = FWW^*W = F(I - F)W = 0$, so that $W^*F = 0$, we have $U^*FU = (V^* + W^*)F(V + W) = V^*FV = V^*(VV^*)V = E$. Also $V^*W = (V^*VV^*)(WW^*W) = (V^*F)(I - F)W = 0$, and hence $W^*V = 0$, so that $U^*U = (V^* + W^*)(V + W) = E + (I - E) = I$. Similarly, $VW^* = WV^* = 0$ and $UU^* = I$.

§8. Topological Properties of *-Isomorphisms

Let $\mathfrak{A}$ be a *-algebra of operators with $I \in \mathfrak{A}$. A *state* of $\mathfrak{A}$ is a complex linear functional f on $\mathfrak{A}$ such that $f(I) = 1$ and $f(A) \geqslant 0$ when $A \geqslant 0$, $A \in \mathfrak{A}$.

LEMMA 18. *Each state f of a *-algebra $\mathfrak{A}$ defines a semidefinite inner product on $\mathfrak{A}$ by the formula*

$$[A, B]_f = f(B^*A),$$

satisfying the Schwarz Inequality

$$|[A, B]_f|^2 \leqslant f(A^*A) \cdot f(B^*B).$$

The proof of this lemma follows the usual pattern (see e.g. [9; pp. 216–217]) and will be omitted.

COROLLARY 16. *Each state is norm continuous, and in fact, $||f|| = 1$, for f a state.*

Proof. For $A \in \mathfrak{A}$ we have $-||A^*A|| \cdot I \leqslant A^*A \leqslant ||A^*A|| \cdot I$, so $f(A^*A) \leqslant ||A^*A||$. Also $||A^*A|| = ||A||^2$ and $|f(A)|^2 = |[A, I]_f|^2 \leqslant f(A^*A) \leqslant ||A||^2$ by Lemma 18, so that $||f|| \leqslant 1$. Since $f(I) = 1$, $||f|| = 1$.

A state (or positive linear functional) f on a von Neumann algebra $\mathfrak{A}$ is *completely additive* if $f(E) = \Sigma f(E_n)$, for each orthogonal family $\{E_n\}$ of projections in $\mathfrak{A}$ with $E = \text{LUB } E_n$.

Note. The infinite sum $\Sigma f(E_n)$ is defined by

$$\Sigma f(E_n) = \lim_{\substack{\varphi \text{ finite}}} \sum_{n \in \varphi} f(E_n)$$

(the limit is taken over the net of finite partial sums, indexed by the finite subsets φ of the index set of $\{E_n\}$, directed by inclusion).

Example. For $||x|| = 1$, the state $A \mapsto (Ax|x)$ is completely additive.

LEMMA 19. *Let $E \neq 0$ be a projection in a von Neumann algebra $\mathfrak{A}$, and let f and g be two completely additive positive linear functionals (i.e., unnormalized states) on $\mathfrak{A}$ with $f(E) \leq g(E)$. Then there is a projection $0 \neq F \leq E$ in $\mathfrak{A}$ with $f(G) \leq g(G)$ for all projections $G \in \mathfrak{A}$ with $0 \neq G \leq F$.*

Proof. Let $\{E_n\}$ be a maximal orthogonal family of projections in $\mathfrak{A}$ with $E_n \leq E$ and $f(E_n) > g(E_n)$, and set $E_0 = \text{LUB } E_n$. Then

$$f(E_0) = \Sigma f(E_n) > \Sigma g(E_n) = g(E_0).$$

Put $F = E - E_0$. Then $0 \neq F \leq E$ and if G is a projection in $\mathfrak{A}$ with $0 < G \leq F$, then $f(G) \leq g(G)$, for otherwise $\{E_n\}$ would fail to be maximal.

Notation. For a subset $\mathfrak{A} \subset \mathfrak{B}(\mathcal{H})$, we write

$$\mathfrak{A}^+ = \{A \in \mathfrak{A}: A \geq 0\}.$$

THEOREM 10. *A state on a von Neumann algebra is completely additive if and only if it is strongly continuous on the unit ball.*

Proof. Let f be a completely additive state on a von Neumann algebra $\mathfrak{A}$ and let $E \neq 0$ be a projection in $\mathfrak{A}$. Choose a vector x so that $f(E) \leq (Ex|x)$ and by Lemma 19 choose a projection $0 < F \leq E$

in $\mathcal{A}$ so that $f(G) \leqslant (Gx|x)$ for all projections $G \in \mathcal{A}$ with $0 < G \leqslant F$. For $A \in \mathcal{A}^+$ with $A = FAF$, the Spectral Theorem shows that $f(A) \leqslant (Ax|x)$. For we may approximate A in the algebra $F\mathcal{A}F$ (a von Neumann algebra whose identity is F) by a linear combination $\sum_{i=1}^{n} a_i E_i$ of commuting projections $E_i \in F\mathcal{A}F$ with each $a_i \geqslant 0$, in such a way that $0 \leqslant A - \sum_{i=1}^{n} a_i E_i \leqslant \epsilon \cdot I$ (where $\epsilon > 0$ is prescribed in advance). But then $f(A) \leqslant f(\sum_{i=1}^{n} a_i E_i) + \epsilon \cdot I(I) = \sum_{i=1}^{n} a_i f(E_i) + \epsilon \leqslant \sum_{i=1}^{n} a_i (E_i x|x) + \epsilon = ((\sum_{i=1}^{n} a_i E_i)x|x) + \epsilon \leqslant (Ax|x) + \epsilon$. The arbitrariness of $\epsilon > 0$ then implies $f(A) \leqslant (Ax|x)$, for all $A \in \mathcal{A}^+$ such that $A = FAF$.

Now by Lemma 18, we have $|f(AF)|^2 = |[AF, I]_f|^2 \leqslant f(FA^*AF) \cdot f(I) \leqslant (FA^*AFx|x) = ||AFx||^2$, for any $A \in \mathcal{A}$

This shows that the mapping $A \mapsto f(AF)$ is strongly continuous. Now let $\{P_n\}$ be an orthogonal family of projections in $\mathcal{A}$ maximal with respect to the condition that each of the maps $A \mapsto f(AP_n)$ is strongly continuous. Let $P = \text{LUB } P_n$. If $P < I$, then by the first part of the proof we can find a projection $0 < F \leqslant I - P$, $F \in \mathcal{A}$, for which $A \mapsto f(AF)$ is strongly continuous, contradicting maximality of the P_n's. Thus $P = I$. Now let $P_\varphi = \sum_{n \in \varphi} P_n$, where φ runs over the finite subsets of the index set of the family $\{P_n\}$. Then $f(I) = f(P) = \Sigma f(P_n) = \lim f(P_\varphi)$. For $A \in \mathcal{A}$ with $||A|| \leqslant 1$ we have $A^*A \leqslant I$ and by Lemma 18, $|f(A(I - P_\varphi))|^2 \leqslant f(A^*A) \cdot f(I - P_\varphi) \leqslant f(I) \cdot f(I - P_\varphi) \to 0$ as φ increases, so the state f is the uniform limit of the maps $A \mapsto f(AP_\varphi)$ on the unit ball. Thus f is strongly continuous on the unit ball.

The converse is an easy consequence of Vigier's Theorem (§1, Lemma 1).

COROLLARY 17. *The two conditions in Theorem 10 are each equivalent to f being normal, i.e., $f(A) = \text{LUB } f(A_n)$ for each directed family $A_n \in \mathcal{A}^+$ with $A = \text{LUB } A_n$.*

Note. $A_n \uparrow A$ strongly and weakly by Vigier's Theorem.

THEOREM 11. *Let $\mathcal{A}$ be a von Neumann algebra and let $\mathcal{S} \subset \mathcal{A}$ be any complex linear subspace. Then the weak and strong closures of $\mathcal{S}$ coincide.*

Proof. Suppose that $\mathcal{S}$ is strongly closed. We show that $\mathcal{S}$ is also weakly closed. We may assume, moreover, that $\mathcal{A} = \mathcal{B}(\mathcal{H})$, since $\mathcal{A}$ is weakly closed.

Write $\mathcal{H} = \mathcal{H}_1 \oplus \ldots \oplus \mathcal{H}_n$ and let U_i be an isometry with $U_i \colon \mathcal{H} \cong \mathcal{H}_i$ (we assume that dim $\mathcal{H} = \infty$, since otherwise all linear space topologies on $\mathcal{B}(\mathcal{H})$ coalesce).

Take an operator A in the weak closure of $\mathcal{S}$, let $\epsilon > 0$ and $x \in \mathcal{H}$ be given, and write $x_i = U_i U_i^* x$ so that $x = x_1 + \ldots + x_n$. For $B \in \mathcal{B}(\mathcal{H})$ define $x_B = U_1 B x_1 + \ldots + U_n B x_n$. Then $B \mapsto x_B$ is a linear map and $\mathfrak{M} = \{x_S \colon S \in \mathcal{S}\}$ is a linear subspace.

If $y \in \mathfrak{M}^\perp$, then $(x_S | y) = 0$, for all $S \in \mathcal{S}$. Now there is an operator $B \in \mathcal{S}$ with $|((A - B)x_i | U_i^* y)| < \epsilon \cdot n^{-1}$, for each i, so that $|(x_A | y)| = |(x_{A-B} | y)| = |(\sum_{i=1}^n U_i (A - B) x_i | y)| \leq \sum_{i=1}^n |((A - B) x_i | U_i^* y)| < \epsilon$. Hence $(x_A | y) = 0$ and therefore $x_A \in \mathfrak{M}^{\perp \perp}$.

Thus we can find $z \in \mathfrak{M}(x = x_S$, for some $S \in \mathcal{S})$ with $||x_A - x_S|| < \epsilon \cdot n^{-1/2}$. But then $\epsilon^2 > n \cdot ||x_A - x_S||^2 = n \cdot ||\sum_{i=1}^n U_i (A - S) x_i||^2 = n \cdot \sum_{i=1}^n ||U_i (A - S) x_i||^2 = n \cdot \sum_{i=1}^n ||(A - S) x_i||^2 \geq (\sum_{i=1}^n ||(A - S) x_i||)^2 \geq ||\sum_{i=1}^n (A - S) x_i||^2 = ||(A - S) x||^2$, so that $||(A - S) x|| < \epsilon$ (*Note.* In the above estimate, we used two facts: (i) For $i \neq k$, ran(U_i) $\perp$ ran(U_k); and (ii) If $a_i \geq 0$, $i = 1, 2, \ldots, n$, then $(\sum_{i=1}^n a_i) \leq n \cdot (\sum_{i=1}^n a_i^2)$. The latter is just the Schwarz Inequality).

Since $\epsilon > 0$ and $x \in \mathcal{H}$ were arbitrary, $A \in \mathcal{S}$ as required, since $\mathcal{S}$ is strongly closed.

COROLLARY 18. *A linear functional f on a von Neumann algebra is weakly continuous if and only if it is strongly continuous.*

Proof. Strong and weak closure of the null space of f amount to the same thing.

COROLLARY 19. *The weak and strong closures of any convex set of operators coincide.*

Proof. Both topologies are locally convex since they are defined by seminorms:

Strong. $A \mapsto ||Ax||$.
Weak. $A \mapsto |(Ax|y)|$ (or $|(Ax|x)|$).

But in a locally convex space, the closure of a convex set is the intersection of all closed half-spaces (of continuous linear functionals) containing it.

COROLLARY 20. *Let $\mathcal{A}$ be a von Neumann algebra, $\mathcal{A}_1$ its unit ball and f a complex linear functional. Then $f|\mathcal{A}_1$ is strongly continuous if and only if $f|\mathcal{A}_1$ is weakly continuous.*

Proof. The restriction $f|\mathcal{A}_1$ of f to $\mathcal{A}_1$ is strongly (resp. weakly) continuous if and only if $f^{-1}(K) \cap \mathcal{A}_1$ is strongly (resp. weakly) closed, for every closed convex set K of complex numbers. The result now follows from Corollary 19.

A map $\Phi: \mathcal{A} \to \mathcal{B}$ between von Neumann algebras is *positive* if $\Phi(\mathcal{A}^+) \subset \mathcal{B}^+$. A positive linear map $\Phi: \mathcal{A} \to \mathcal{B}$ between von Neumann algebras is *completely additive* if $\Phi(E) = \Sigma \Phi(E_n)$, for each orthogonal family $\{E_n\}$ of projections in $\mathcal{A}$ with $E = \text{LUB } E_n$ (the sum on the right is interpreted as before, with convergence in the strong operator topology). Note that if Φ is a *-algebra homomorphism, then $\Phi(I) = P$ is a projection in $\mathcal{B}$, so on replacing $\mathcal{B}$ by $P\mathcal{B}P$, we can assume that $\Phi(I) = I$. A *-algebra homomorphism is automatically positive, since $\mathcal{A}^+ = \{A \in \mathcal{A}: A = X^*X, \text{ some } X \in \mathcal{A}\}$.

THEOREM 12. *Let $\mathfrak{A}$ and $\mathfrak{B}$ be von Neumann algebras and let $\Phi: \mathfrak{A} \to \mathfrak{B}$ be a completely additive *-algebra homomorphism. Then Φ is weakly (resp. strongly) continuous on the unit ball of $\mathfrak{A}$, if $\mathfrak{B}$ has the weak (resp. strong) operator topology.*

Proof. For each completely additive state f on $\mathfrak{B}$, $f \circ \Phi$ is a completely additive state on $\mathfrak{A}$, hence is strongly and weakly continuous on $\mathfrak{A}_1 = \{A \in \mathfrak{A}: ||A|| \leq 1\}$ by Theorem 10 and Corollary 20.

For any unit vector x, let ω_x be the *vector state* defined by: $\omega_x(T) = (Tx|x)$. Taking $f = \omega_x|\mathfrak{B}$ in the above, we see that if $\{T_n\}$ is a net with $T_n \to 0$ weakly and $||T_n|| \leq 1$, then $(\Phi(T_n)x|x) = (\omega_x \circ \Phi)(T_n) \to 0$ for each unit vector x, so that $\Phi(T_n) \to 0$ weakly, proving the first statement.

Let $\{T_n\}$ be a net in $\mathfrak{A}$ with $T_n \to 0$ strongly and $||T_n|| \leq 1$. Now $||(T_n^*T_n)^{1/2}x||^2 = ((T_n^*T_n)^{1/2}x|(T_n^*T_n)^{1/2}x) = (T_n^*T_n x|x) = ||T_n x||^2 \to 0$. Hence $(T_n^*T_n)^{1/2} \to 0$ strongly, and since $||T_n^*T_n x|| \leq ||(T_n^*T_n)^{1/2}|| \cdot ||(T_n^*T_n)^{1/2}x|| \to 0$ (note that $||(T_n^*T_n)^{1/2}|| = ||T_n^*T_n||^{1/2} = ||T_n|| \leq 1$) we also have $T_n^*T_n \to 0$ strongly. Thus $||(\Phi T_n)x||^2 = ((\Phi T_n)^*(\Phi T_n)x|x) = \omega_x[(\Phi T_n)^*(\Phi T_n)] = (\omega_x \circ \Phi)(T_n^*T_n) \to 0$, so that $\Phi T_n \to 0$ strongly, as claimed.

COROLLARY 21. *The conclusions of Theorem 12 hold if and only if Φ is normal, i.e., preserves LUB's of directed families in $\mathfrak{A}^+$.*

COROLLARY 22. *Let $\mathfrak{A}$ and $\mathfrak{B}$ be von Neumann algebras and let $\Phi: \mathfrak{A} \cong \mathfrak{B}$ be a *-algebra isomorphism. Then Φ is strongly (resp. weakly) continuous on the unit ball of $\mathfrak{A}$, if $\mathfrak{B}$ has the strong (resp. weak) operator topology.*

Proof. Φ is an order isomorphism of the s.a. part of $\mathfrak{A}$ with the s.a. part of $\mathfrak{B}$, hence Φ is normal. Now apply Corollary 21 and Theorem 12.

THEOREM 13. *The norm unit ball of a von Neumann algebra is compact in the weak operator topology.*

Proof. It is enough to prove this for the algebra $\mathcal{A} = \mathcal{B}(\mathcal{H})$. For $x \in \mathcal{H}$, let

$$K_x = \{y \in \mathcal{H}: ||y|| \leqslant ||x||\}.$$

By reflexivity of $\mathcal{H}$, each K_x is weakly compact in $\mathcal{H}$. Let $K = \Pi K_x \subset \mathcal{H}^{\mathcal{H}}$ (product over all $x \in \mathcal{H}$). By the Tychonoff Theorem, K is compact. Furthermore, for $T \in \mathcal{A}_1 = \{T \in \mathcal{A}: ||T|| \leqslant 1\}$, $||Tx|| \leqslant ||x||$, so $\mathcal{A}_1 \subset K$. It is therefore enough to show that $\mathcal{A}_1$ is closed in the product topology of $\mathcal{H}^{\mathcal{H}}$ (each copy of $\mathcal{H}$ being given the weak Hilbert space topology), since on $\mathcal{A}_1$, this is the weak operator topology.

Note that $K = \{T \in \mathcal{H}^{\mathcal{H}}: ||Tx|| \leqslant ||x||,$ all $x \in \mathcal{H}\}$. Now fix a scalar a and two vectors $x, y \in \mathcal{H}$. Then the mapping

$$T \mapsto T(ax + y) - a(Tx) - Ty,$$

call it $\pi_{a, x, y}$, is continuous from $\mathcal{H}^{\mathcal{H}}$ (with the product topology described above) to $\mathcal{H}$ (with the weak Hilbert space topology). For if $\{T_n\}$ is a net with $T_n \to T$ (product), then clearly $(\pi_{a, x, y}(T_n)|z) \to (\pi_{a, x, y}(T)|z)$, for all $z \in \mathcal{H}$. Hence the inverse image $\pi_{a, x, y}^{-1}(0)$ of zero is closed in $\mathcal{H}^{\mathcal{H}}$. If $\mathcal{L} = \{T \in \mathcal{H}^{\mathcal{H}}: T$ is linear$\}$, then

$$\mathcal{L} = \cap \pi_{a, x, y}^{-1}(0),$$

where the intersection is taken over all scalars a and all $x, y \in \mathcal{H}$. This shows that $\mathcal{L}$ is closed, and since $\mathcal{A}_1 = \mathcal{L} \cap K$, the latter is too, proving the theorem.

THE KAPLANSKY DENSITY THEOREM. *Let $\mathcal{A}$ and $\mathcal{B}$ be *-algebras of operators on $\mathcal{H}$ with $\mathcal{A} \subset \mathcal{B}$ and suppose that $\mathcal{A}$ is* dense in $\mathcal{B}$ in the strong operator topology. Then $\mathcal{C}_1$ is strongly dense in $\mathcal{B}_1$.

Exercises. (1) The converse is trivial. (2) With $\mathcal{C}$ and $\mathcal{B}$ as above, $M_2(\mathcal{C})$ is strongly dense in $M_2(\mathcal{B})$, on $\mathcal{H} \oplus \mathcal{H}$.

Proof (Dixmier [1]). We can assume that $\mathcal{C}$ and $\mathcal{B}$ are norm closed (i.e., are C^*-algebras), since $\mathcal{C}_1$ is norm dense in the unit ball of the norm closure of $\mathcal{C}$.

Let $\mathcal{S} = \{A \in \mathcal{C}: A^* = A\}$ and $\mathcal{T} = \{B \in \mathcal{B}: B^* = B\}$. For $T \in \mathcal{T}$, T belongs to the weak closure of $\mathcal{C}$ by Theorem 11. But the mapping $A \mapsto \frac{1}{2}(A + A^*)$ is weakly continuous (since $A \mapsto A^*$ is), leaves T fixed and takes $\mathcal{C}$ onto $\mathcal{S}$. Hence T belongs to the weak closure of $\mathcal{S}$, and therefore to the strong closure of $\mathcal{S}$ by Theorem 11 again, since $\mathcal{S}$ is convex.

Now assume $||T|| \leqslant 1$. The function $x \mapsto 2x(1 + x^2)^{-1}$ is strictly increasing on the interval $[-1, 1]$ and takes all values between -1 and 1. Since $\mathcal{B}$ is a C^*-algebra, there is an operator $T' \in \mathcal{T}$ with $T = 2T'(I + T'^2)^{-1}$ (apply the inverse of the above function to T to get T').

Let S' be any operator in $\mathcal{S}$ and put $S = 2S'(I + S'^2)^{-1}$. Since $|2x(1 + x^2)^{-1}| \leqslant 1$ for all $x \in [|1, 1|]$, we have $||S|| \leqslant 1$, and $\frac{1}{2}(S - T) = (I + S'^2)^{-1}[S'(I + T'^2) - (I + S'^2)T'](I + T'^2)^{-1} = (I + S'^2)^{-1}(S' - T')(I + T'^2)^{-1} + (I + S'^2)^{-1} S'(T' - S')T'(I + T'^2)^{-1} = (I + S'^2)^{-1}(S' - T')(I + T'^2)^{-1} + \frac{1}{4}S(T' - S')T$. When S' tends strongly to T', S tends strongly to T, since $||(I + S'^2)^{-1}|| \leqslant 1$ and $||S|| \leqslant 1$ (note that the map $(S, T) \mapsto ST$ is strongly continuous if we keep S in the unit ball). Hence $\mathcal{S}_1$ is strongly dense in $\mathcal{T}_1$.

Now take $T \in \mathcal{B}$, $||T|| \leqslant 1$. Then

$$\begin{pmatrix} 0 & T \\ T^* & 0 \end{pmatrix} \in M_2(\mathcal{B})$$

is self-adjoint and since

$$\left\|\begin{pmatrix} 0 & T \\ T^* & 0 \end{pmatrix}\right\|^2 = \left\|\begin{pmatrix} 0 & T \\ T^* & 0 \end{pmatrix}^2\right\| = \left\|\begin{pmatrix} TT^* & 0 \\ 0 & T^*T \end{pmatrix}\right\|$$
$$= \max(||TT^*||, ||T^*T||) = ||T||^2,$$

this operator matrix is also a contraction. Hence there is a net

$$\begin{pmatrix} S_n^{11} & S_n^{12} \\ S_n^{21} & S_n^{22} \end{pmatrix} \in M_2(\mathcal{C})$$

of self-adjoint contractions converging strongly to

$$\begin{pmatrix} 0 & T \\ T^* & 0 \end{pmatrix}.$$

But then $S_n^{12} \to T$ strongly with $||S_n^{12}|| \leq 1$ and $S_n^{12} \in \mathcal{C}$, and the proof is complete.

Remark. If $\mathcal{C}_1$ is strongly closed, so is $\mathcal{C}$. (*Proof.* Take $T \in \mathcal{C}''$. Since $\mathcal{C}$ is strongly dense in $\mathcal{C}''$ by the Double Commutant Theorem of von Neumann, $\mathcal{C}_1$ is strongly dense in $(\mathcal{C}_1'')$ by the Kaplansky Density Theorem. Thus $||T||^{-1} \cdot T$ is the strong limit of a net $T_n \in \mathcal{C}_1$. But then $||T||^{-1} \cdot T \in \mathcal{C}_1$, so $T \in \mathcal{C}$.)

THEOREM 14. *Let $\mathcal{C}$ and $\mathcal{B}$ be von Neumann algebras and let $\Phi: \mathcal{C} \to \mathcal{B}$ be a completely additive *-algebra homomorphism. Then $\Phi(\mathcal{C})$ is a von Neumann algebra.*

Proof. By one of the remarks before Theorem 12, we can assume $\Phi(I) = I$. Let $\mathcal{J} = \Phi^{-1}(0)$. Then $\mathcal{J}$ is a (two-sided) *-ideal whose unit ball is weakly and strongly closed by Theorem 12. Hence $\mathcal{J}$ is a von Neumann algebra by the remark before the theorem. Let E be its identity. For any projection $G \in \mathcal{C}$, let $S = 2G - I$ (s.a., unitary).

Then SES is a projection in $\mathcal{J}$ and hence $SES \leqslant E$. But $S^2 = I$, so $E = S(SES)S < SES$ and $SE = ES$. Thus E is central in $\mathcal{C}$, and also $\mathcal{J} = \mathcal{C}E$.

On the von Neumann algebra $\mathcal{R} = \mathcal{C}(I - E)$, Φ is one-to-one. Furthermore, Φ is isometric on $\mathcal{R}$. For if $R \in \mathcal{R}$, $||R^*R|| = ||R||^2$ and $R^*R \leqslant ||R||^2 \cdot I$ (§0, 27) so that $(\Phi R)^*(\Phi R) \leqslant ||R||^2 \cdot \Phi(I) = ||R||^2 \cdot I$. Hence $||\Phi(R)|| \leqslant ||R||$.

Now suppose $S \in \mathcal{R}^+$. If p is any real polynomial, $\Phi(p(S)) = p(\Phi(S))$ and by continuity and the Weierstrass Theorem, this relation holds for all continuous functions f. But if $||\Phi(S)|| < ||S||$, we could find an f with $f(S) \neq 0$, but $f(\Phi(S)) = 0$, so $\Phi(f(S)) = 0$, a contradiction. Thus Φ is an isometry.

Now the unit ball $\mathcal{C}_1$ of $\mathcal{C}$ is compact in the weak operator topology (Theorem 13) and $\Phi(\mathcal{C}_1) = (\Phi\mathcal{C})_1$, the unit ball of $\Phi(\mathcal{C})$, by the isometric nature of Φ. Since $\Phi|\mathcal{C}_1$ is weakly continuous (Theorem 12), $(\Phi\mathcal{C})_1$ is weakly compact, and hence strongly closed. The remark before the theorem then implies that $\Phi(\mathcal{C})$ is a von Neumann algebra.

INVARIANCE OF GENERATION. *Let $\mathcal{C}$ and $\mathcal{B}$ be von Neumann algebras and let $\Phi: \mathcal{C} \cong \mathcal{B}$ be a *-algebra isomorphism. Suppose that the family $\{A_n\}$ generates $\mathcal{C}$ as a von Neumann algebra and set $B_n = \Phi(A_n)$, for each n. Then the family $\{B_n\}$ generates $\mathcal{B}$ as a von Neumann algebra.*

Proof. If $\mathcal{M}$ (resp. $\mathcal{N}$) is the C^*-algebra generated by $\{A_n\}$ (resp. $\{B_n\}$), then $\mathcal{M}$ is strongly dense in $\mathcal{C}$, so by the Kaplansky Density Theorem, $\mathcal{M}_1$ is strongly dense in $\mathcal{C}_1$. But then $\mathcal{N}_1 = \Phi(\mathcal{M}_1)$ is strongly dense in $\mathcal{B}_1$ by Corollary 22, so $\mathcal{N}$ is strongly dense in $\mathcal{B}$ by the (trivial) converse of the Kaplansky Density Theorem.

§9. Algebras on Separable Spaces

Certain technical features of the theory become more manageable if we assume that the underlying Hilbert space is separable. We now explore a few consequences of this assumption.

Although the following result makes no use of separability, our only application of it will be in that context.

PROPOSITION 9. *Let $\mathfrak{A}$ be a von Neumann algebra acting on the Hilbert space $\mathcal{H}$, and suppose that $\mathfrak{X}$ is a cyclic set of vectors for $\mathfrak{A}'$. Let $\{T_n\}$ be a net of operators in $\mathfrak{A}$ with $\sup_n ||T_n|| = M < \infty$. Then $T_n \to T \in \mathfrak{A}$ strongly if and only if $||(T - T_n)x|| \to 0$, as $n\uparrow$, for each $x \in \mathfrak{X}$.*

·*Proof.* Let $\mathfrak{M}$ be the set of all vectors $x \in \mathcal{H}$ for which $||(T - T_n)x|| \to 0$. If λ is a scalar and $x, y \in \mathfrak{M}$, then $||(T - T_n)(\lambda x + y)|| \leq |\lambda| \cdot ||(T - T_n)x|| + ||(T - T_n)y|| \to 0$. Hence $\mathfrak{M}$ is a linear subspace of $\mathcal{H}$. We show that $\mathfrak{M}$ is closed.

To this end, take $x \in [\mathfrak{M}]$ (the closure of $\mathfrak{M}$), and let $\epsilon > 0$ be given. Choose $y \in \mathfrak{M}$ so that $||x - y|| < \epsilon \cdot (M + ||T||)^{-1}$. Then $||(T - T_n)(x - y)|| < ||T - T_n|| \cdot ||x - y|| < \epsilon$, and therefore $|\ ||(T - T_n)x|| - ||(T - T_n)y||\ | < \epsilon$. Since $\lim_n ||(T - T_n)y|| = 0$, this implies that $\limsup_n ||(T - T_n)x|| \leq \epsilon$, so, in fact, $\lim_n ||(T - T_n)x|| = 0$ and $x \in \mathfrak{M}$. This shows that $\mathfrak{M}$ is closed.

Now if $A' \in \mathfrak{A}'$, we have $||(T - T_n)A'x|| = ||A'(T - T_n)x|| \leq ||A'|| \cdot ||(T - T_n)x|| \to 0$, for each $x \in \mathfrak{X}$. Hence $\mathfrak{A}'\mathfrak{X} \subset \mathfrak{M}$ and since $\mathfrak{M}$

is closed, we have $\mathcal{H} = [\mathcal{A}'\mathfrak{X}] \subset \mathfrak{M}$ by the cyclicity hypothesis, so that $\mathfrak{M} = \mathcal{H}$ and therefore $T_n \to T$ strongly.

The reverse implication is clear.

LEMMA 20. *Every von Neumann algebra $\mathcal{A}$ on a separable Hilbert space $\mathcal{H}$ has a countable separating family $\{x_n\}_{n=1}^{\infty}$ (which can be chosen to be orthonormal).*

Proof. Let $\{x_n\}$ be a family of unit vectors maximal (Zorn) with respect to the condition $[\mathcal{A}'x_m] \perp [\mathcal{A}'x_n]$ if $m \neq n$. By separability of $\mathcal{H}$, $\{x_n\}_{n=1}^{\infty}$ is at most countable (being an orthonormal set).

Let E_n denote the projection onto $[\mathcal{A}'x_n]$ and put $E = \text{LUB } E_n$. Note that each $E_n \in \mathcal{A}$ by Proposition 5(3). We assert that $E = I$. If not, consider $0 \neq x_0 \in \text{ran}(I - E)$. Then $0 \neq [\mathcal{A}'x_0] \subset \text{ran}(I - E)$ and the projection onto $[\mathcal{A}'x_0]$ belongs to $\mathcal{A}$, by Proposition 5(3,5). Hence $\{x_n, x_0\}$ is a strictly larger family with the above properties, contradicting maximality of $\{x_n\}$. Thus $E = I$.

To see that $\{x_n\}_{n=0}^{\infty}$ is separating for $\mathcal{A}$, suppose that $A \in \mathcal{A}$ satisfies $Ax_n = 0$ for $n = 1, 2, \ldots$. If $A' \in \mathcal{A}'$, then $AA'x_n = A'A_n = 0$ for all n, so that A annihilates the (dense) linear subspace spanned by $\bigcup_{n=1}^{\infty} \mathcal{A}'x_n$, and is therefore the zero operator by continuity.

COROLLARY 23. *Every von Neumann algebra $\mathcal{A}$ on a separable Hilbert space $\mathcal{H}$ possesses a countable orthonormal sequence $\{x_n\}_{n=1}^{\infty}$ which is cyclic for $\mathcal{A}$.*

Proof. By Lemma 20, $\mathcal{A}'$ has a countable (orthonormal) separating family $\{x_n\}_{n=1}^{\infty}$. The remark after Proposition 5 shows that a subset $\mathfrak{X} \subset \mathcal{H}$ is cyclic for $\mathcal{A}$ if and only if $\mathfrak{X}$ is separating for $\mathcal{A}'$. Taking $\mathfrak{X} = \{x_n\}_{n=1}^{\infty}$, we obtain the desired result.

COROLLARY 24. *An abelian von Neumann algebra $\mathcal{A}$ on a separable Hilbert space $\mathcal{H}$ has a separating vector.*

ALGEBRAS ON SEPARABLE SPACES 63

Proof. Apply Lemma 20 to obtain an orthonormal separating sequence $\{x_n\}_{n=1}^{\infty}$ for $\mathcal{A}$.

Let $x_0 = x_1 + \frac{1}{2} x_2 \; \frac{1}{3} x_3 \; \ldots$. Then

$$\|x_0\|^2 = \sum_{n=1}^{\infty} \left\|\frac{1}{n} x_n\right\|^2 = \sum_{n=1}^{\infty} \frac{1}{n^2} = \frac{\pi^2}{6} < \infty,$$

so $x_0 \in \mathcal{H}$.

Finally, suppose $A \in \mathcal{A}$ and $Ax_0 = 0$. Then if E_n denotes the projection onto $[\mathcal{A}'x_n]$, $E_n \in \mathcal{A}$ and $Ax_n = nAE_n x_0 = nE_n Ax_0 = 0$ for each n. Since $\{x_n\}_{n=1}^{\infty}$ is a separating family for $\mathcal{A}$, $A = 0$.

THEOREM 15. *Let $\mathcal{A}$ be a von Neumann algebra acting on a separable Hilbert space $\mathcal{H}$, and let $\mathcal{A}_1 = \{A \in \mathcal{A} : \|A\| \leq 1\}$ be the norm unit ball. Then in the strong operator topology $\mathcal{A}_1$ is separably metrizable.*

Proof. Choose (Lemma 20) an orthonormal sequence $\{x_n\}_{n=1}^{\infty}$ in $\mathcal{H}$ which is separating for $\mathcal{A}$ (hence cyclic for $\mathcal{A}'$—see the remark after Proposition 5).

By Proposition 9, strong operator convergence of nets in $\mathcal{A}_1$ is the same as pointwise convergence (in the norm topology of $\mathcal{H}$) on the countable set $\{x_n\}_{n=1}^{\infty}$. Thus the map $T \mapsto (Tx_1, Tx_2, \ldots)$ is a homeomorphism of $\mathcal{A}_1$ (strong operator topology) with a subspace of the countable product

$$\mathcal{H}^{\omega} = \mathcal{H} \times \mathcal{H} \times \ldots \quad \text{(product topology)}$$

of $\mathcal{H}$ with itself, $\mathcal{H}$ having its usual norm topology.

However, this supspace of $\mathcal{H}^{\omega}$ is metrizable, and in fact $\mathcal{A}_1$ can be given the metric

$$d(A, B) = \sum_{n=1}^{\infty} 2^{-n} \|(A - B)x_n\|.$$

Since $\mathcal{H}$ is separable, so is the image of $\mathcal{A}_1$ in $\mathcal{H}^\omega$, as is $\mathcal{A}_1$ itself.

COROLLARY 25. *Every von Neumann algebra acting on a separable Hilbert space is countably generated (as a von Neumann algebra).*

Proof. By Theorem 15, $\mathcal{A}$ contains a countable sequence $\{A_n\}_{n=1}^{\infty}$ which is dense (*sequentially*, in fact) in $\mathcal{A}_1$ in the strong operator topology.

But the set $\mathcal{G} = \{I, A_n, A_n^*\}_{n=1}^{\infty}$ is also countable and clearly $\mathcal{G}'' = \mathcal{A}$.

Exercise. If $\mathcal{A}$ acts on a separable space, the weak operator topology on $\mathcal{A}_1$ is separably metrizable (*Hints.* Imitate the proof of Proposition 9 and define the metric by

$$d(A, B) = \sum_{m,n=1}^{\infty} 2^{-(m+n)} |(Ax_m|x_n) - (Bx_m|x_n)|).$$

Problem. Is every von Neumann algebra acting on a separable Hilbert space finitely generated? (An affirmative answer to this would imply that each such algebra is *singly* generated, as will be shown in the next section.)

THEOREM 16. *Let $\mathcal{A}$ be a von Neumann algebra. Then $\mathcal{A}$ is *-isomorphic to a von Neumann algebra acting on a separable Hilbert space if and only if the following two conditions hold:*

(1) *$\mathcal{A}$ is countably generated, and*

(2) *$\mathcal{A}$ is countably decomposable (i.e., every orthogonal family of projections in $\mathcal{A}$ is countable).*

Proof. Suppose that the von Neumann algebra $\mathcal{A}$ is *-isomorphic to a separably acting von Neumann algebra. Condition (1) then obtains from Corollary 25 and the Invariance of Generation Theorem of §8. Condition (2) obviously holds.

For the converse, suppose that $\mathcal{C}$ is generated by the sequence $\{A_n\}_{n=1}^{\infty}$ and let $\{x_n\}_{n=1}^{\infty}$ be an orthonormal sequence which is separating for $\mathcal{C}$ (copy the proof of Lemma 20, replacing the separability hypothesis by condition (2)).

Let E' be the projection onto $[\mathcal{C} x_n : n = 1, 2, \ldots]$. Then $E' \in \mathcal{C}'$ by Proposition 5(3). Furthermore, since $\mathcal{C}$ contains a countable subset which is dense in the strong operator topology (namely, rational linear combinations of the *-semigroup generated by $\{A_n\}_{n=1}^{\infty}$), the subspace ran(E') is easily seen to be separable.

By Lemma 10, ran($C_{E'}$) = $[\mathcal{C}' \operatorname{ran}(E')] = \mathcal{H}$, since the remark after Proposition 5 implies that the sequence $\{x_n\}_{n=1}^{\infty}$ is cyclic for $\mathcal{C}'$.

By Lemma 13 and exercise (2) before Corollary 5, we have *-algebra isomorphisms

$$\mathcal{C} \cong \mathcal{C} E' \cong \mathcal{C}_{E'},$$

and the last algebra acts separably. The proof is complete.

Exercise. Let $(X, \mathfrak{M}, \mu)$ be any measure space for which $L^{\infty}(X, \mu)$ is the dual of $L^1(X, \mu)$. If $\mathcal{C} = L^{\infty}$ acts by multiplication on L^2, show that the weak *-topology on $\mathcal{C}$ (viewed as the dual of L^1) is the weak operator topology on $\mathcal{C}$ (viewed as operators on L^2), hence that $\mathcal{C}$ is a von Neumann algebra in this representation.

Examples and Remarks. (1) Let $\mathcal{C}$ be a countably generated von Neumann algebra whose center $\mathcal{Z} = \mathcal{C} \cap \mathcal{C}'$ is countably decomposable. Then $\mathcal{C}$ is *-isomorphic to a separably acting von Neumann algebra (*Proof.* Choose a separating vector x_0 for $\mathcal{Z}$, imitating the proofs of Lemma 20 and Corollary 24, replacing the separability hypothesis by countable decomposability. Let E' be the projection onto $[\mathcal{C} x_0]$.

Then $E' \in \mathfrak{A}'$ by Proposition 5(3). As in the proof of Theorem 16, ran(E') is separable. If G is a central projection with $GE' = 0$, then $GAx_0 = 0$, for each $A \in \mathfrak{A}$. In particular, $Gx_0 = 0$, so $G = 0$. Hence E is faithful by the first exercise after Theorem 3. The last paragraph of the proof of Theorem 16 now applies and completes the argument.)

(2) Let $\mathcal{H} = l^2[0,1]$, where $[0,1]$ is the discrete unit interval with counting measure. Let $\mathfrak{A} = l^\infty[0,1]$ act on $\mathcal{H}$ by multiplication. Then $\mathfrak{A}$ is an abelian von Neumann algebra, since the weak *-topology on $\mathfrak{A}$ is the weak operator topology when $\mathfrak{A}$ is viewed as operators on $\mathcal{H}$ (see exercise above). Further, the identity function $(x \mapsto x)$ generates $\mathfrak{A}$ as a von Neumann algebra. But $\mathfrak{A}$ is *not* countably decomposable (consider characteristic functions of points), so by Theorem 16, $\mathfrak{A}$ cannot be faithfully represented to act separably.

(3) Let $(X, \mathfrak{M}, \mu)$ be a finite measure space for which $L^2(X,\mu)$ is *not* separable (*Exercise.* Find one such.) Let $\mathfrak{A} = L^\infty(X,\mu)$ act by multiplication on $L^2(X,\mu)$. Then 1 is a cyclic vector, so $\mathfrak{A}$ is a maximal abelian von Neumann algebra by Theorem 5 (the same comment about the w^*-topology and the weak operator topology made in (2) above applies here). If $\mathfrak{A}$ were countably generated, then $L^2(X,\mu) = [\mathfrak{A} \cdot 1]$ would be separable, as in the proof of Theorem 16, a contradiction. Hence $\mathfrak{A}$ has no separable representation which is faithful.

§10. Generators

We now take up the question of which separably acting von Neumann algebras can be generated by a single operator. Along the way, we shall also consider small sets of generators having special properties.

The following simple but useful result will be employed both in this section and the next.

RICKART'S LEMMA ([18]). *Let X be a compact Hausdorff space, and suppose that the algebra $C(X)$ of all continuous complex functions on X is generated as a Banach algebra (sup norm) by a countable family of idempotents. Then $C(X)$ is generated by a single real function.*

Proof. Let $\{e_n\}_{n=1}^{\infty}$ be the sequence of idempotents generating $C(X)$. Define, for $x \in X$,

$$h(x) = \sum_{n=1}^{\infty} 3^{-n}(2e_n(x)-1).$$

Since $2e_n(x)-1 = \pm 1$ for each $x \in X$, the above series converges uniformly and defines a real function in $C(X)$. Because the e_n's generate $C(X)$ uniformly, they must separate points of X. Hence if $x \neq y$ are in X, there is a smallest integer k for which $e_k(x) \neq e_k(y)$.

But then $|h(x) - h(y)| = 2|\sum_{n=k}^{\infty} 3^{-n}(e_n(x) - e_n(y))| \geq 2 \cdot 3^{-k} - 2 \cdot \sum_{n=k+1}^{\infty} 3^{-n} = 3^{-k}$. This proves that h separates X. Since h is real, the Stone-Weierstrass Theorem implies that h generates $C(X)$.

We now give what appears to be a new proof of one of the earliest results on generators.

THEOREM 17 (von Neumann [15]). *Let $\mathcal{A}$ be an abelian von Neumann algebra acting on a separable Hilbert space. Then there is a single self-adjoint operator $H \in \mathcal{A}$ which generates $\mathcal{A}$ as a von Neumann algebra.*

Proof. By Corollary 25, $\mathcal{A}$ has a countable set of generators, all in the unit ball. Breaking each of these into its real and imaginary parts, we see that $\mathcal{A}$ is generated by a sequence $\{A_n\}_{n=1}^{\infty}$ of self-adjoint contractions.

Now let $A_n = \int \lambda \, dE_\lambda^{(n)}$ be the spectral resolution of A_n. Then

$$E_\lambda^{(n)} = \underset{\mu > \lambda}{\text{GLB}} \, E_\mu^{(n)} \quad (\mu \text{ rational}),$$

and $E_\mu^{(n)} \downarrow E_\lambda^{(n)}$ strongly by Lemma 1. Hence $\mathcal{A}$ is generated by a countable family of projections, say $\{E_n\}_{n=1}^{\infty}$.

Let $\mathcal{C}$ be the (abelian) C^*-algebra generated by the E_n's. By the Gelfand Representation Theorem, $\mathcal{C} \cong C(X)$, where X is the (compact) maximal ideal space of $\mathcal{C}$. Let e_n denote the idempotent in $C(X)$ corresponding to $E_n \in \mathcal{C}$ under this isometric $*$-isomorphism. Defining $h \in C(X)$ from the e_n's as in Rickart's Lemma, and letting $H \in \mathcal{C}$ correspond to h, we see that the (s.a.) operator H generates $\mathcal{C}$ as a C^*-algebra, and since $\mathcal{C}$ is strongly dense in $\mathcal{A}$, we have $\mathcal{A} = \{H\}''$ by the Double Commutant Theorem.

The next few results deal with algebras which are singly, doubly or triply generated by special operators. The proofs are arranged so that they depend on the following simple observation [4].

THEOREM 18. *Let $\mathfrak{A}$ be a von Neumann algebra acting on the Hilbert space $\mathcal{H}$. If A generates $\mathfrak{A}$ as a von Neumann algebra, then*

$$\begin{pmatrix} 0 & I \\ 0 & 0 \end{pmatrix} \text{ and } \begin{pmatrix} 0 & A \\ 0 & 0 \end{pmatrix}$$

generate $M_2(\mathfrak{A})$ as a von Neumann algebra acting on $\mathcal{H} \oplus \mathcal{H}$ (each of these operators has square equal to zero).

Proof. Let $\mathfrak{M}$ denote the von Neumann algebra on $\mathcal{H} \oplus \mathcal{H}$ generated by

$$\begin{pmatrix} 0 & I \\ 0 & 0 \end{pmatrix} \text{ and } \begin{pmatrix} 0 & A \\ 0 & 0 \end{pmatrix}.$$

Let $\mathcal{G}$ be the set consisting of these operators and their adjoints. Then $\mathfrak{M} = \mathcal{G}''$ and $\mathfrak{M}' = \mathcal{G}'$. We first determine the s.a. operators in $\mathfrak{M}'$. Take $X, Y, Z \in \mathcal{B}(\mathcal{H})$ with X, Z s.a. The typical s.a. operator on $\mathcal{H} \oplus \mathcal{H}$ has the form

$$\begin{pmatrix} X & Y \\ Y^* & Z \end{pmatrix}.$$

Assume that this operator commutes with

$$\begin{pmatrix} 0 & I \\ 0 & 0 \end{pmatrix} \text{ and } \begin{pmatrix} 0 & A \\ 0 & 0 \end{pmatrix}.$$

Then $Y = 0$, $X = Z$ and X commutes with A and A^*. Hence (see Remark 2(i) after Proposition 2),

$$\mathfrak{M}' = \left\{ \begin{pmatrix} A' & 0 \\ 0 & A' \end{pmatrix} : A' \in \mathfrak{A}' \right\}.$$

Finally, a similar computation shows that

$$\mathfrak{M} = \left\{ \begin{pmatrix} A' & 0 \\ 0 & A' \end{pmatrix} : A' \in \mathfrak{A}' \right\}'$$

$$= \left\{ \begin{pmatrix} X & Y \\ Z & W \end{pmatrix} : X, Y, Z, W \in \mathfrak{A}'' = \mathfrak{A} \right\} = M_2(\mathfrak{A}),$$

proving the theorem.

COROLLARY 26 ([4]). *If A generates $\mathfrak{A}$ as a von Neumann algebra on $\mathcal{H}$, then the two idempotents*

$$\begin{pmatrix} I & I \\ 0 & 0 \end{pmatrix} \text{ and } \begin{pmatrix} I & A \\ 0 & 0 \end{pmatrix}$$

generate $M_2(\mathfrak{A})$ as a von Neumann algebra on $\mathcal{H} \oplus \mathcal{H}$.

Proof. Let $\mathfrak{M}$ be the von Neumann algebra generated by

$$\begin{pmatrix} I & I \\ 0 & 0 \end{pmatrix} \text{ and } \begin{pmatrix} I & A \\ 0 & 0 \end{pmatrix}.$$

Then

$$\begin{pmatrix} I & 0 \\ 0 & 0 \end{pmatrix} = \frac{1}{2} \begin{pmatrix} I & I \\ 0 & 0 \end{pmatrix} \begin{pmatrix} I & 0 \\ I & 0 \end{pmatrix},$$

$$\begin{pmatrix} 0 & I \\ 0 & 0 \end{pmatrix} = \begin{pmatrix} I & I \\ 0 & 0 \end{pmatrix} - \begin{pmatrix} I & 0 \\ 0 & 0 \end{pmatrix}$$

and

$$\begin{pmatrix} 0 & A \\ 0 & 0 \end{pmatrix} = \begin{pmatrix} I & A \\ 0 & 0 \end{pmatrix} - \begin{pmatrix} I & 0 \\ 0 & 0 \end{pmatrix}$$

all belong to $\mathfrak{M}$, so $\mathfrak{M} = M_2(\mathfrak{A})$ by Theorem 18.

Remark. If A generates $\mathfrak{A}$, so does $a(A + \lambda I)$, where $a, \lambda > 0$. Hence, by translating and scaling, we may assume that A is invertible and that $||A|| < 1$.

COROLLARY 27 ([16]). *If A is an invertible strict contraction which generates $\mathfrak{A}$ on $\mathcal{H}$, then the partial isometry*

$$\begin{pmatrix} A & 0 \\ T & 0 \end{pmatrix}, \qquad \text{where } T = (I - A^*A)^{1/2},$$

generates $M_2(\mathfrak{A})$ on $\mathcal{H} \oplus \mathcal{H}$.

Proof. Since $r(A^*A) = ||A^*A|| = ||A||^2 < 1$ ($r(\cdot)$ = the spectral radius of $(\cdot)$), $1 \notin \text{sp}(A^*A)$ so T^2, and hence T, is invertible. Also

$$\begin{pmatrix} I & 0 \\ 0 & 0 \end{pmatrix} = \begin{pmatrix} A^* & T^* \\ 0 & 0 \end{pmatrix} \begin{pmatrix} A & 0 \\ T & 0 \end{pmatrix},$$

so that

$$\begin{pmatrix} A & 0 \\ T & 0 \end{pmatrix}$$

is a partial isometry. Let X, Y, $Z \in \mathcal{B}(\mathcal{H})$ with X, Z s.a., and assume that the s.a. operator

$$\begin{pmatrix} X & Y \\ Y^* & Z \end{pmatrix}$$

commutes with

$$\begin{pmatrix} A & 0 \\ T & 0 \end{pmatrix}.$$

Then $AY = 0$, so $Y = A^{-1}(AY) = 0$. Further, $AX = XA$ and $TX = ZT$, so $Z = TXT^{-1}$. But X commutes with A and A^*, and hence with $I - A^*A$ and T, which implies $X = Z$. But clearly

$$\begin{pmatrix} X & 0 \\ 0 & X \end{pmatrix}$$

commutes with

$$\begin{pmatrix} 0 & I \\ 0 & 0 \end{pmatrix} \text{ and } \begin{pmatrix} 0 & A \\ 0 & 0 \end{pmatrix}$$

if X commutes with A. Thus if $\mathfrak{A}$ is the von Neumann algebra generated by

$$\begin{pmatrix} A & 0 \\ T & 0 \end{pmatrix},$$

we have

$$\begin{pmatrix} 0 & I \\ 0 & 0 \end{pmatrix}, \begin{pmatrix} 0 & A \\ 0 & 0 \end{pmatrix} \in \left\{ \begin{pmatrix} A & 0 \\ T & 0 \end{pmatrix}, \begin{pmatrix} A^* & T^* \\ 0 & 0 \end{pmatrix} \right\}'' = \mathfrak{M}.$$

By Theorem 18, $\mathfrak{M} = M_2(\mathfrak{A})$.

Remark.

$$\begin{pmatrix} A & 0 \\ T & 0 \end{pmatrix}$$

is the Halmos unitary dilation

$$\begin{pmatrix} A & S \\ T & -A^* \end{pmatrix}$$

of A ([9]; Solution 177) with the second column deleted.

COROLLARY 28 ([20]). *If the invertible strict contraction A generates $\mathfrak{A}$ on $\mathcal{H}$, then the three projections*

$$E = \begin{pmatrix} I & 0 \\ 0 & 0 \end{pmatrix}, \quad F = \frac{1}{2}\begin{pmatrix} I & I \\ I & I \end{pmatrix}, \quad G = \begin{pmatrix} AA^* & SA \\ A^*S & T^2 \end{pmatrix}$$

(where $S = (I - AA^*)^{1/2}$ and $T = (I - A^*A)^{1/2}$) generate $M_2(\mathfrak{A})$ on $\mathcal{H} \oplus \mathcal{H}$.

Proof. Let $\mathfrak{M}$ be the von Neumann algebra generated by E, F and G. Then

$$\begin{pmatrix} 0 & I \\ 0 & 0 \end{pmatrix} = 2EF - E \in \mathfrak{M}$$

and

$$\begin{pmatrix} 0 & 0 \\ I & 0 \end{pmatrix} = \begin{pmatrix} 0 & I \\ 0 & 0 \end{pmatrix}^* \in \mathfrak{M},$$

so that the symmetry

$$\begin{pmatrix} 0 & I \\ I & 0 \end{pmatrix} = \begin{pmatrix} 0 & I \\ 0 & 0 \end{pmatrix} + \begin{pmatrix} 0 & 0 \\ I & 0 \end{pmatrix}$$

belongs to $\mathfrak{M}$. Also

$$\begin{pmatrix} SA & AA^* \\ T^2 & A^*S \end{pmatrix} = G \cdot \begin{pmatrix} 0 & I \\ I & 0 \end{pmatrix} \in \mathfrak{M}$$

and hence

$$\begin{pmatrix} SA & 0 \\ 0 & 0 \end{pmatrix} = E \cdot \begin{pmatrix} SA & AA^* \\ T^2 & A^*S \end{pmatrix} \cdot E \in \mathfrak{M}.$$

But

$$\begin{pmatrix} I - AA^* & 0 \\ 0 & 0 \end{pmatrix} = E - EGE \in \mathfrak{M}$$

so

$$\begin{pmatrix} S & 0 \\ 0 & 0 \end{pmatrix} \in \mathfrak{M} \text{ and } \begin{pmatrix} S^{-1} & 0 \\ 0 & 0 \end{pmatrix} \in \mathfrak{M}$$

since S^{-1} is a norm limit of polynomials in S. But then

$$\begin{pmatrix} A & 0 \\ 0 & 0 \end{pmatrix} = \begin{pmatrix} S^{-1} & 0 \\ 0 & 0 \end{pmatrix} \begin{pmatrix} SA & 0 \\ 0 & 0 \end{pmatrix} \in \mathfrak{M}$$

and

$$\begin{pmatrix} 0 & A \\ 0 & 0 \end{pmatrix} = \begin{pmatrix} A & 0 \\ 0 & 0 \end{pmatrix} \begin{pmatrix} 0 & I \\ 0 & 0 \end{pmatrix} \in \mathfrak{M}.$$

By Theorem 18, $\mathfrak{M} = M_2(\mathfrak{A})$.

Remark. The three projections of Corollary 28 are unitarily equivalent in $M_2(\mathfrak{A})$. For

$$U = 2^{-1/2}\begin{pmatrix} I & I \\ I & -I \end{pmatrix} \text{ and } V = \begin{pmatrix} A & S \\ T & -A^* \end{pmatrix}$$

are unitary in $M_2(\mathfrak{A})$ with $E = V^*GV$.

COROLLARY 29 ([20]). *If the invertible strict contraction A generates $\mathfrak{A}$ on $\mathcal{H}$, then the two unitary operators*

$$W = \begin{pmatrix} I & 0 \\ 0 & -I \end{pmatrix} \text{ and } V = \begin{pmatrix} A & S \\ T & -A^* \end{pmatrix}$$

(where $S = (I - AA^*)^{1/2}$ and $T = (I - A^*A)^{1/2}$) generate $M_2(\mathfrak{A})$ on $\mathcal{H} \oplus \mathcal{H}$.

Proof. Let $\mathfrak{M}$ be the von Neumann algebra generated by W and V. Then

$$\begin{pmatrix} I & 0 \\ 0 & 0 \end{pmatrix} = \frac{1}{2}\left(I + W\right) \in \mathfrak{M}$$

and

$$\begin{pmatrix} A & 0 \\ T & 0 \end{pmatrix} = V \cdot \begin{pmatrix} I & 0 \\ 0 & 0 \end{pmatrix} \in \mathfrak{M}.$$

By Corollary 27, $\mathfrak{M} = M_2(\mathfrak{A})$.

Remark. The first unitary in Corollary 29 is actually a symmetry.

The remaining results of this section are due to W. Wogen [24].

Notation. $(A, B, \ldots)$ denotes the von Neumann algebra generated by the family $\{A, B, \ldots\}$ of operators.

LEMMA 21. *Suppose a von Neumann algebra $\mathcal{A}$ on the Hilbert space $\mathcal{H}$ is generated by n operators $A_1, A_2, \ldots, A_n$ ($n \geqslant 2$). Then $M_2(\mathcal{A})$ is generated by the $n+1$ operators*

$$\begin{pmatrix} A_1 & 0 \\ 0 & 0 \end{pmatrix}, \begin{pmatrix} A_2 & 0 \\ 0 & 0 \end{pmatrix}, \ldots, \begin{pmatrix} A_n & 0 \\ 0 & 0 \end{pmatrix}, \begin{pmatrix} 0 & I \\ 0 & 0 \end{pmatrix}$$

on $\mathcal{H} \oplus \mathcal{H}$.

The proof is an easy exercise and will be omitted.

LEMMA 22. *Suppose a von Neumann algebra $\mathcal{A}$ on $\mathcal{H}$ is generated by n operators $A_1, \ldots, A_n$ ($n \geqslant 2$), one of which, say A_1, is normal. Then $M_2(\mathcal{A})$ is generated by $n-1$ operators on $\mathcal{H} \oplus \mathcal{H}$.*

Proof. By the remark before Corollary 27, we may assume that $A_1, \ldots, A_n$ are invertible strict contractions. Put

$$B_i = \begin{pmatrix} A_i & 0 \\ 0 & 0 \end{pmatrix}$$

for $i = 1, 2, \ldots, n-1$ and let

$$U = \begin{pmatrix} A_n & S_n \\ T_n & -A_n^* \end{pmatrix},$$

where $S_n = (I - A_n A_n^*)^{1/2}$ and $T = (I - A_n^* A_n)^{1/2}$. Then U is unitary (the Halmos dilation of A_n) and B_1 is normal, so $\mathcal{R}(U)$ and $\mathcal{R}(B_1)$ have s.a. generators, say G and H, respectively, by Theorem 17. Then $C = G + iH$ generates $\mathcal{R}(U, B_1)$ and therefore $\mathcal{R}(B_1, \ldots, B_{n-1}, U) = \mathcal{R}(C, B_2, \ldots, B_{n-1})$. It will be enough to show that $M_2(\mathcal{A}) = \mathcal{R}(B_1, \ldots, B_{n-1}, U)$.

Let $\mathcal{R} = \mathcal{R}(B_1, \ldots, B_{n-1}, U)$. Then

$$\begin{pmatrix} I & 0 \\ 0 & 0 \end{pmatrix} = \begin{pmatrix} A_1 & 0 \\ 0 & 0 \end{pmatrix} \begin{pmatrix} A_1^{-1} & 0 \\ 0 & 0 \end{pmatrix} \in \mathcal{R},$$

so that

$$X = \begin{pmatrix} A_n & 0 \\ T_n & 0 \end{pmatrix} = U \cdot \begin{pmatrix} I & 0 \\ 0 & 0 \end{pmatrix} \in \mathcal{R}.$$

By Corollary 27, $M_2(\mathcal{R}(A_n)) = \mathcal{R}(X)$, so

$$\begin{pmatrix} A_1 & 0 \\ 0 & 0 \end{pmatrix}, \ldots, \begin{pmatrix} A_n & 0 \\ 0 & 0 \end{pmatrix}, \begin{pmatrix} 0 & I \\ 0 & 0 \end{pmatrix} \in \mathcal{R},$$

and $\mathcal{R} = M_2(\mathcal{A})$ by Lemma 21.

COROLLARY 30. *Let* $\mathcal{A} = \mathcal{R}(A_1, \ldots, A_n)$, $n \geqslant 3$, *where the first three operators* A_1, A_2, A_3 *are normal. Then* $M_2(\mathcal{A})$ *is generated by* $n - 2$ *operators.*

Proof. The von Neumann algebras $\mathcal{R}(A_2)$ and $\mathcal{R}(A_3)$ are abelian, so $\mathcal{R}(A_2, A_3)$ has a single generator B, as in the proof of Lemma 22. But then $\mathcal{A} = \mathcal{R}(A_1, B, A_4, \ldots, A_n)$, with A_1 normal, so by Lemma 22, $M_2(\mathcal{A})$ is generated by $n - 2$ operators.

LEMMA 23. *Let* $\mathcal{A}$ *be a von Neumann algebra acting on the Hilbert space* $\mathcal{H}$, *and suppose that* $\mathcal{A} = \mathcal{R}(A_1, \ldots, A_n)$ *with* $n \geqslant 2$. *Then* $M_2(\mathcal{A})$ *is generated by* $n + 1$ *unitary operators on* $\mathcal{H} \oplus \mathcal{H}$.

Proof. As usual, we suppose that $A_1, \ldots, A_n$ are invertible strict contractions. For $i = 1, \ldots, n$ let

$$U_i = \begin{pmatrix} A_i & S_i \\ T_i & -A_i^* \end{pmatrix} \text{ and } W = \begin{pmatrix} I & 0 \\ 0 & -I \end{pmatrix},$$

where $S_i = (I - A_i A_i^*)^{1/2}$ and $T_i = (I - A_i^* A_i)^{1/2}$ (the U_i's are unitary and W is a symmetry). Then

$$\begin{pmatrix} A_1 & 0 \\ T_1 & 0 \end{pmatrix} = U_1 \cdot \begin{pmatrix} I & 0 \\ 0 & 0 \end{pmatrix} \in \mathcal{R}(U_1, W),$$

since

$$\begin{pmatrix} I & 0 \\ 0 & -I \end{pmatrix} = \tfrac{1}{2}(I + W).$$

But then $\Re(U_1, W) = M_2(\Re(A_1))$ by Corollary 27, and hence we have

$$\begin{pmatrix} A_1 & 0 \\ 0 & 0 \end{pmatrix}, \ldots, \begin{pmatrix} A_n & 0 \\ 0 & 0 \end{pmatrix}, \begin{pmatrix} 0 & I \\ 0 & 0 \end{pmatrix} \in \Re(U_1, \ldots, U_n, W),$$

so that $\Re(U_1, \ldots, U_n, W) = M_2(\mathfrak{A})$ by Lemma 21.

Roughly speaking, the next theorem reduces the single generator problem to showing that a countably generated algebra can be finitely generated.

THEOREM 19. *Let $\mathfrak{A}$ be a finitely generated von Neumann algebra which is *-isomorphic to the ring $M(\mathfrak{A})$ of all 2×2 matrices over itself. Then $\mathfrak{A}$ is singly generated.*

Proof. Assume that $\mathfrak{A}$ is generated by $(\geqslant 2)$ operators. Then $M_2(\mathfrak{A}) \cong \mathfrak{A}$ is generated by $n + 1 (\geqslant 3)$ unitary operators by Lemma 23. Thus $M_2(\mathfrak{A}) \cong \mathfrak{A}$ is generated by $(n + 1) - 2 = n - 1$ operators by Corollary 30. It follows by repetition that $\mathfrak{A}$ is singly generated.

Exercise. If $\mathcal{H}$ is a separable Hilbert space, the algebra $\mathcal{B}(\mathcal{H})$ is singly generated. The generators of $\mathcal{B}(\mathcal{H})$ are precisely the *irreducible* operators, i.e., operators whose only reducing subspaces are $\{0\}$ and $\mathcal{H}$. One such operator is the *simple unilateral shift* S defined on an orthonormal basis $\{e_n\}_{n=0}^{\infty}$ by setting $Se_n = e_{n+1}$ (S extends by linearity and continuity to an isometry on $\mathcal{H}$). If $A \in \{S\}'$ and if $a_{ij} = (Ae_j | e_i)$, for $i, j = 0, 1, 2, \ldots$, then the matrix (a_{ij}) of A satisfies: (i) $a_{ij} = 0$ for $i < j$, and (ii) $a_{i+1, j+1} = a_{ij}$ for $i \geqslant j$. Formulate similar condition for the matrix of an operator $A \in \{S^*\}'$, and show that $\{S, S^*\}' = \{S\}' \cap \{S^*\}' = \{\lambda I\}$. Hence $\{S, S^*\}'' = \mathcal{B}(\mathcal{H})$.

The following result solves the generator problem in all but the II_1 case (see [3] for further results).

WOGEN'S THEOREM. *Every properly infinite von Neumann algebra acting on a separable Hilbert space is singly generated.*

Proof. Let $\mathcal{A}$ be such an algebra, and let $\{A_n\}_{n=1}^{\infty}$ generate $\mathcal{A}$, with each $\|A_n\| \leq 1$ (Corollary 25). Put

$$A = \begin{pmatrix} A_1 & & & \\ & A_2 & & 0 \\ & & A_3 & \\ & 0 & & \ddots \end{pmatrix} \in M_{\infty}(\mathcal{A}),$$

where $M_{\infty}(\mathcal{A})$ is the algebra of $\aleph_0 \times \aleph_0$ matrices with entries from $\mathcal{A}$ which act boundedly on $\Sigma_{n=1}^{\infty} \oplus \mathcal{H}$ (where $\mathcal{A}$ acts on $\mathcal{H}$).

We denote by $M_{\infty}(\mathbf{C} \cdot I)$ the subalgebra of $M_{\infty}(\mathcal{A})$ whose entries are scalar multiples of I. It is easy to see that $M_{\infty}(\mathbf{C} \cdot I)$ is *-isomorphic with $\mathcal{B}(\mathcal{H})$. But by the above exercise, $\mathcal{B}(\mathcal{H})$ is singly generated, as is $M_{\infty}(\mathbf{C} \cdot I)$, say by an operator B, by the Invariance of Generation Theorem.

We assert that $\mathcal{R}(A, B) = M_{\infty}(\mathcal{A})$. For suppose the s.a. operator C commutes with B. Let $E_{ij} \in M_{\infty}(\mathbf{C} \cdot I)$ be the matrix with I (the identity operator on $\mathcal{H}$) in the (i, j)-th position and zeros elsewhere. Since $M_{\infty}(\mathbf{C} \cdot I) = \{B, B^*\}''$, C commutes with each E_{ij}. A familiar matrix computation arises from equating the (m, n)-th entries of $E_{ij}C$ and CE_{ij} and gives the relation $\delta_{im}C_{jn} = \delta_{jn}C_{mi}$, from which it follows that (i) $C_{ij} = 0$ if $i \neq j$, and (ii) the diagonal entries $C_{ii} = D$ are constant and self-adjoint (because C is). Hence

$$C = \begin{pmatrix} D & & & \\ & D & & 0 \\ & & D & \\ & & & \ddots \\ 0 & & & \end{pmatrix}.$$

Also, if C commutes with A, then D commutes with each A_n, so $D \in \mathcal{C}'$. It follows easily that

$$\mathcal{R}(A, B)' = \left\{ \begin{pmatrix} A' & & & \\ & A' & & 0 \\ & & A' & \\ & & & \ddots \\ 0 & & & \end{pmatrix} : A' \in \mathcal{C}' \right\},$$

and therefore that

$$\mathcal{R}(A, B)'' = M_\infty(\mathcal{C}).$$

But by Theorem 19, $\mathcal{R}(A, B)$ is singly generated, as is therefore, the algebra

$$\mathcal{C} \cong M_\infty(\mathcal{C})$$

(the last isomorphism comes from Corollary 14).

Exercise. Let $\mathcal{C}$ be a von Neumann algebra on $\mathcal{H}$ with a single generator A. Then

$$T = \begin{pmatrix} 0 & I & A \\ 0 & 0 & I \\ 0 & 0 & 0 \end{pmatrix}$$

generates $M_3(\mathcal{C})$ as a von Neumann algebra on $\mathcal{H} \oplus \mathcal{H} \oplus \mathcal{H}$, and $T^3 = 0$. Hence by Wogen's Theorem and Corollary 14, any separably acting properly infinite von Neumann algebra has a generator T satisfying $T^3 = 0$ (Pearcy and Ringrose).

§11. Hyperfinite Algebras

In this section we consider an important class of von Neumann algebras which arise naturally in representation theory. The construction of these algebras occurs in two stages. First we produce a class of C*-algebras, called UHF algebras, via a process which involves building an increasing family of 'compatible' matrix units. The second step involves finding appropriate representations of UHF algebras.

More specifically, let $\{p_n\}_{n=1}^{\infty}$ be a sequence of positive integers with $2 \leq p_1 < p_2 < p_3 < \ldots$. A *uniformly hyperfinite (UHF) algebra of type* $\{p_n\}$ is a C*-algebra $\mathfrak{A}$ (i.e., a norm closed *-subalgebra of $\mathcal{B}(\mathcal{H})$) containing a sequence $\mathfrak{M}_n \subset \mathfrak{A}$ of von Neumann factors $(n = 1, 2, \ldots)$ such that

(1) $\mathfrak{M}_n$ is of type I_{p_n} (i.e., $\mathfrak{M}_n$ is *-isomorphic to the algebra $M_{p_n}(\mathbf{C})$ of all $p_n \times p_n$ complex matrices).

(2) $I \in \mathfrak{M}_n$, for each n.

(3) $\mathfrak{M}_n \subset \mathfrak{M}_{n+1}$, for each n.

(4) $\mathfrak{A}$ is the norm closure of $\cup_{n=1}^{\infty} \mathfrak{M}_n$.

Remark. UHF algebras exist (*Proof.* Let $\mathcal{H}$ be any infinite dimensional Hilbert space, and let E and F be projections with $\dim E = \dim F$, $EF = 0$ and $E + F = I$. Choose a partial isometry U from $\text{ran}(E)$ to $\text{ran}(F)$. Then the complex linear span of E, F, U and U^* is a factor $\mathfrak{M}_1$ of type I_2. Next write $E = G + H$ as the sum of projections G and H, where dim G = dim H and $GH = 0$. Let V be a partial isometry from ran(G) to ran(H). Then the *algebra* generated by $\mathfrak{M}_1$, G, H, V and V^* is a factor $\mathfrak{M}_2$ of type I_4. Continuing in this way, we get an ascending sequence $\{\mathfrak{M}_n\}$ of factors on $\mathcal{H}$, with $\mathfrak{M}_n$ of type I_{2^n}. The norm closure of the *-subalgebra $\cup_{n=1}^{\infty} \mathfrak{M}_n$ of $\mathfrak{B}(\mathcal{H})$ is a UHF algebra $\mathfrak{A}$ of type $\{2^n\}$. In most of what follows, we shall be mainly concerned with this algebra).

For an $n \times n$ matrix $A = (a_{ij})$ we define the *(normalized) trace* of A by

$$\text{tr}(A) = \frac{1}{n} \sum_{i=1}^{n} a_{ii}.$$

Exercise. tr is a state on $M_n(\mathbf{C})$.

Remark. Let $\mathfrak{M} = M_n(\mathbf{C})$ and let f be a linear functional on $\mathfrak{M}$ such that: (1) $f(I) = 1$, and (2) $f(AB) = f(BA)$, for all $A, B \in \mathfrak{M}$. Then $f = \text{tr}$. (*Proof.* Let $\{E_{ij}\}_{i,j=1}^{n}$ be matrix units, i.e., $\sum_{i=1}^{n} E_{ii} = I$, $E_{ij}E_{kl} = \delta_{jk}E_{il}$ and $E_{ij}^* = E_{ji}$. Then if $i \neq j$, $f(E_{ij}) = f(E_{ii}E_{ij}) = f(E_{ij}E_{ii}) = f(0) = 0$. Also $f(E_{ii}) = f(E_{ij}E_{ji}) = f(E_{ji}E_{ij}) = f(E_{jj})$, so $f(E_{ii}) = n^{-1}$.)

LEMMA 24. *Let $\mathfrak{M}$ and $\mathfrak{N}$ be type I_m and I_n subfactors, respectively, of a C^*-algebra $\mathfrak{A}$, with $I \in \mathfrak{M} \subset \mathfrak{N}$. Then $m|n$.*

Proof. On the projections in $\mathfrak{M}$, the trace has values $\{0, m^{-1}, 2m^{-1}, \ldots, (m-1)m^{-1}, 1\}$, while the trace on $\mathfrak{N}$ has properties (1) and (2) of the above remark on $\mathfrak{M}$, hence $\text{tr}_{\mathfrak{N}}|\mathfrak{M} = \text{tr}_{\mathfrak{M}}$. On the projection in $\mathfrak{N}$, the trace has values $\{0, n^{-1}, 2n^{-1}, \ldots, (n-1)n^{-1}, 1\}$. Thus $m^{-1} = pn^{-1}$ or $n = pm$, for a suitable integer p.

Remark. As a consequence of the last lemma, the p_n's defining the type of a UHF algebra satisfy $p_n | p_{n+1}$.

HYPERFINITE ALGEBRAS

Example. Let $\mathfrak{M}_2 = M_2(M_2(\mathbf{C})) \cong M_4(\mathbf{C})$, and let

$$\mathfrak{M}_1 = \left\{ \begin{pmatrix} \alpha I & \beta I \\ \gamma I & \delta I \end{pmatrix} : \alpha, \beta, \gamma, \delta \in \mathbf{C} \right\},$$

where

$$I = \begin{pmatrix} 1 & 0 \\ 0 & 1 \end{pmatrix}.$$

Then $\mathfrak{N}_1 = \mathfrak{M}_1 \cap \mathfrak{M}_0' = \mathfrak{M}_1$, where

$$\mathfrak{M}_0 = \left\{ \begin{pmatrix} \alpha I & 0 \\ 0 & \alpha I \end{pmatrix} : \alpha \in \mathbf{C} \right\}.$$

Let

$$\begin{pmatrix} A & B \\ C & D \end{pmatrix} \in \mathfrak{N}_2 = \mathfrak{M}_2 \cap \mathfrak{M}_1'$$

where $A, B, C, D \in M_2(\mathbf{C})$. Then

$$\begin{pmatrix} A & B \\ C & D \end{pmatrix}$$

commutes with

$$\begin{pmatrix} 0 & I \\ 0 & 0 \end{pmatrix} \text{ and } \begin{pmatrix} 0 & 0 \\ I & 0 \end{pmatrix} \quad (\in \mathfrak{M}_1),$$

which implies $A = D$ and $B = C = 0$. Hence

$$\mathfrak{N}_2 = \left\{ \begin{pmatrix} A & 0 \\ 0 & A \end{pmatrix} : A \in M_2(\mathbf{C}) \right\}.$$

Clearly $\mathfrak{N}_1$ and $\mathfrak{N}_2$ are commuting subfactors of $\mathfrak{M}_2$ (of type I_2) which generate $\mathfrak{M}_2$ as an algebra.

Exercise. Let $\mathfrak{A}$ be a UHF algebra and let $\{\mathfrak{M}_n\}_{n=1}^{\infty}$ be its defining sequence of type I_{p_n} subfactors. Let $\mathfrak{M}_0 = \{aI: a \text{ complex}\}$ and define $\mathfrak{N}_n = \mathfrak{M}_n \cap \mathfrak{M}'_{n-1}$ ($n = 1, 2, \ldots$). Then
(1) $\mathfrak{N}_n$ is a factor of type $p_n \cdot p_{n-1}^{-1}$.
(2) $\mathfrak{N}_m \subset \mathfrak{N}'_n$ if $m \neq n$.
(3) $\cup_{n=1}^{m} \mathfrak{N}_n$ generates $\mathfrak{M}_m$ (algebraically).
(4) $\cup_{n=1}^{\infty} \mathfrak{N}_n$ generates $\mathfrak{A}$ as a C^*-algebra.

Remark. A decomposition into mutually commuting factors which generate $\mathfrak{A}$ (such as the above) is called a *factorization* of $\mathfrak{A}$ (see [17]).

Exercise. $M_n(\mathbf{C})$ has a single generator (as a $*$-algebra, even), e.g., the matrix

$$\begin{pmatrix} 1 & 1 & 0 & 0 & & \\ 0 & 2 & 1 & 0 & & 0 \\ 0 & 0 & 3 & 1 & & \\ & & \cdot & \cdot & & \\ & & & \cdot & \cdot & \\ & 0 & & & n-1 & 1 \\ & & & & 0 & n \end{pmatrix}.$$

THEOREM 20 ([22]). *Every UHF algebra is singly generated as a C^*-algebra.*

Proof. Let $\mathfrak{A}$ be a UHF algebra and let $\{\mathfrak{N}_n\}_{n=1}^{\infty}$ be a sequence of mutually commuting finite type I factors which generate $\mathfrak{A}$ (i.e., $\mathfrak{A}$ is the norm closure of the algebra generated by $\cup_{n=1}^{\infty} \mathfrak{N}_n$—note that the latter set is $*$-closed. The $\mathfrak{N}_n$'s can be chosen as in the exercise before last.)

Now each $\mathfrak{N}_n$ has a single generator (by the preceding exercise); call it G_n. The real part of G_n is a finite real linear combination of mutually commuting projections $\{E_i^{(n)}\}$ in $\mathfrak{N}_n$, and the imaginary part of G_n is a real linear combination of mutually commuting projections $\{F_i^{(n)}\}$ in $\mathfrak{N}_n$. Let $\mathcal{E}$ be the collection of all $E_i^{(n)}$'s, and $\mathcal{F}$ the collection of all $F_i^{(n)}$'s for $n = 1, 2, \ldots$.

Then $\mathcal{E}$ is a countable commuting family of projections, as is $\mathcal{F}$. Let $\mathcal{R}$ (resp. $\mathcal{S}$) be the abelian C^*-algebra generated by $\mathcal{E}$ (resp. $\mathcal{F}$). By Rickart's Lemma (§10), $\mathcal{R}$ (resp. $\mathcal{S}$) has a single s.a. C^*-generator R (resp. S). Put $G = R + iS$. We assert that G generates $\mathfrak{A}$ as a C^*-algebra.

To see this, let $\mathcal{G}$ be the C^*-algebra generated by G. Then $R, S \in \mathcal{G}$, so that $\mathcal{R}, \mathcal{S} \subset \mathcal{G}$. Thus $G_n \in \mathcal{G}$, for each n. But then $\mathfrak{N}_n \subset \mathcal{G}$, for each n, so that $\mathfrak{A} \subset \mathcal{G}$ and therefore $\mathfrak{A} = \mathcal{G}$.

A von Neumann algebra is said to be *hyperfinite* if it is the weak closure of a UHF algebra.

COROLLARY 31. *Every hyperfinite von Neumann algebra is singly generated as a von Neumann algebra.*

Proof. Let $\mathfrak{A}$ be a UHF algebra and let $\{\mathfrak{M}_n\}$ be its defining sequence of type I_{p_n} subfactors. Define a linear functional τ on $\cup_{n=1}^{\infty} \mathfrak{M}_n$ by setting

$$\tau(A) = \mathrm{tr}_n(A)$$

if $A \in \mathfrak{M}_n$ and tr_n denotes the trace on $\mathfrak{M}_n$. Then τ is a state (since each tr_n is) of $\cup_{n=1}^{\infty} \mathfrak{M}_n$. By Corollary 16, τ is norm continuous, hence extends uniquely by continuity to a state τ of $\mathfrak{A}$. We call τ the *trace* of $\mathfrak{A}$.

Remark. $\tau(AB) = \tau(BA)$, for $A, B \in \mathfrak{A}$. (*Proof.* Let $A_n, B_n \in \cup_{n=1}^{\infty} \mathfrak{M}_n$ with $||A - A_n|| \to 0$ and $||B - B_n|| \to 0$. Then $A_n B_n - B_n A_n \to AB - BA$, and since $\tau(A_n B_n - B_n A_n) = 0$ for all n, $\tau(AB - BA) = 0$ by norm continuity of τ.)

We now describe a well-known method of constructing
∗-representations of a C^*-algebra from states. Given a C^*-algebra
$\mathfrak{A}$ and a state f of $\mathfrak{A}$, we can define a semidefinite inner product
(see Lemma 18) on $\mathfrak{A}$ by setting

$$[A, B]_f = f(B^*A).$$

If $\mathfrak{L} = \{A \in \mathfrak{A}: [A, A] = 0\}$, then $\mathfrak{L}$ is a left ideal of $\mathfrak{A}$. For if $A \in \mathfrak{A}$
and $L \in \mathfrak{L}$, we have $[AL, AL] = f(L^*A^*AL) = [L, A^*AL] \leqslant$
$f(L^*L)^{1/2} \cdot f(L^*A^*AA^*AL) = 0$ by the Schwarz Inequality. We
call $\mathfrak{L}$ the *left kernel* of f.

Under the inner product

$$[A + \mathfrak{L}, B + \mathfrak{L}]_f = f(B^*A)$$

induced in $h = \mathfrak{A}/\mathfrak{L}$, h is a pre-Hilbert space. Now $\mathfrak{A}$ acts on the
completion (again denoted by h) as follows. Let $\pi(A)$ denote the
operator (*a priori* unbounded) defined by the formula

$$\pi(A)(B + \mathfrak{L}) = AB + \mathfrak{L}.$$

Then on writing $||A + \mathfrak{L}||_2 = f(A^*A)^{1/2}$ and noting that
$0 \leqslant B^*A^*AB \leqslant ||A||^2 \cdot B^*B$, we have

$$||\pi(A)(B + \mathfrak{L})||_2 = ||AB + \mathfrak{L}||_2 = f(B^*A^*AB)^{1/2}$$
$$\leqslant ||A|| \cdot f(B^*B) = ||A|| \cdot ||B + \mathfrak{L}||_2.$$

Hence $\pi(A)$ is bounded, extends uniquely to the completion h, and
$||\pi(A)|| \leqslant ||A||$. Clearly the mapping $\pi: \mathfrak{A} \to \mathcal{B}(h)$ is an algebra
homomorphism. To see that π preserves adjoints, observe that

$$[\pi(A^*)(B + \mathfrak{L}), C + \mathfrak{L}]_f = [A^*B + \mathfrak{L}, C + \mathfrak{L}]_f$$
$$= f(C^*A^*B) = [B + \mathfrak{L}, AC + \mathfrak{L}]_f$$
$$= [B + \mathfrak{L}, \pi(A)(C + \mathfrak{L})]_f$$

for all B, $C \in \mathfrak{A}$, so that $\pi(A^*) = \pi(A)^*$, for $A \in \mathfrak{A}$.

Finally, $\pi(\mathfrak{A})$ is cyclic on $\hbar$, with cyclic vector $I + \mathfrak{L}$. For

$$[\pi(\mathfrak{A})(I + \mathfrak{L})] = [A + \mathfrak{L}: A \in \mathfrak{A}]$$
$$= \left(||\cdot||\text{-closure of } \frac{\mathfrak{A}}{\mathfrak{L}}\right) = \hbar.$$

Conversely, if $\pi: \mathfrak{A} \to \mathcal{B}(\hbar)$ is a cyclic $*$-representation with cyclic vector x_0, then $f(A) = (\pi(A)x_0 | x_0)$ defines a state of $\mathfrak{A}$.

The next few results deal with the process of factoring a C*-algebra by a closed ideal. The resulting Banach quotient algebra turns out to be representable as a C*-algebra of operators on a suitable Hilbert space, but we shall not need this fact (see [2; 1.8.2 and 2.6.1] for a proof).

LEMMA 25. *Let $\mathfrak{A}$ be a C*-algebra of operators with $I \in \mathfrak{A}$, and let $\mathfrak{J}$ be a norm closed (two-sided) ideal in $\mathfrak{A}$. Then $T^* \in \mathfrak{J}$ whenever $T \in \mathfrak{J}$.*

Proof. Let $T \in \mathfrak{J}$ and put $S = T^*T$. Let $U_n = S(n^{-1} \cdot I + S)^{-1}$ (§0, 9)). Then $[T(I - U_n)]^* [T(I - U_n)] = (I - U_n)S(I - U_n) = S(I + nS)^{-2}$, since $I - U_n = I - S(n^{-1} \cdot I + S)^{-1} = n^{-1}(n^{-1} \cdot I + S)^{-1} = (I + nS)^{-1}$. Also $I \leq I + nS$, hence $0 \leq (I + nS)^{-1} \leq I$ (§0, 22)). Now $||S(I + nS)^{-2}|| \leq ||(I - nS)^{-1}|| \cdot ||S(I + nS)^{-1}|| \leq ||S(I + nS)^{-1}|| \leq n^{-1} \cdot ||I - (I + nS)^{-1}|| \leq n^{-1}$. Thus $||T(I - U_n)||^2 = ||[T(I - U_n)]^*[T(I - U_n)]|| \leq n^{-1}$, and therefore $||T(I - U_n)|| \to 0$. But then $||T^* - U_n T^*|| = ||T - TU_n|| \to 0$, and since $U_n \in \mathfrak{J}$ and $\mathfrak{J}$ is norm closed, $T^* \in \mathfrak{J}$.

COROLLARY 32. *Let $\mathfrak{A}$ be a C*-algebra of operators with $I \in \mathfrak{A}$ and let $\mathfrak{J}$ be a norm closed ideal in $\mathfrak{A}$. Then $\mathcal{B} = \mathfrak{A}/\mathfrak{J}$ is a Banach algebra with involution satisfying $||B^*|| = ||B||$, for each $B \in \mathcal{B}$.*

Proof. The involution $A + \mathcal{J} \mapsto A^* + \mathcal{J}$ is well-defined by Lemma 25, and everything except perhaps the last statement is standard. To see that, notice that since $\mathcal{J} = \mathcal{J}^*$, $||A + \mathcal{J}|| =$ $\inf\{||A + T||: T \in \mathcal{J}\} = \inf\{||A^* + T^*||: T \in \mathcal{J}\} = \inf\{||A^* + T||: T \in \mathcal{J}\} = ||A^* + \mathcal{J}||$.

LEMMA 26. *Let $\mathcal{A}$ be a C*-algebra of operators with $I \in \mathcal{A}$, let $\mathcal{B}$ be a Banach algebra with involution satisfying $||B^*|| = ||B||$, for each $B \in \mathcal{B}$, and let $\pi: \mathcal{A} \cong \mathcal{B}$ be a *-algebra isomorphism of $\mathcal{A}$ onto $\mathcal{B}$. Then $||\pi(A)|| \geq ||A||$, for each $A \in \mathcal{A}$.*

Proof. First note that the existence of $\pi(A)^{-1} \in \mathcal{B}$ implies the existence of $A^{-1} \in \mathcal{A}$, so that $\mathrm{sp}(A) \subset \mathrm{sp}(\pi(A))$. Thus $r(A) \leq r(\pi(A)) \leq ||\pi(A)||$. Finally, $||A||^2 = ||A^*A|| = r(A^*A) \leq ||\pi(A^*A)|| = ||\pi(A)^*\pi(A)|| \leq ||\pi(A^*)|| \cdot ||\pi(A)|| = ||\pi(A)||^2$ (§0, 24) and 27)), proving the assertion.

Exercise. The algebra $M_n(\mathbf{C})$ of all $n \times n$ complex matrices ($2 \leq n < \infty$) is *simple* (i.e., has no ideals except the trivial two).

THEOREM 21 ([5]). *Every UHF algebra is simple, i.e., has no nontrivial ideals.*

Proof. Let $\mathcal{A}$ be the norm closure of an increasing sequence $\mathfrak{M}_1 \subset \mathfrak{M}_2 \subset \ldots \subset \mathfrak{M}_n \subset \ldots$ of factors of type I_{p_n}. Suppose $\mathcal{J}$ is a norm closed ideal of $\mathcal{A}$, $\mathcal{J} \neq \mathcal{A}$. Then $\mathcal{J} \cap \mathfrak{M}_n = 0$ (otherwise $I \in \mathcal{J}$, by the above exercise). Consider the quotient map $\pi: \mathcal{A} \to \mathcal{A}/\mathcal{J}$; then $||\pi(A)|| = \inf\{||A + T||: T \in \mathcal{J}\} \leq ||A||$. Also $\pi|\mathfrak{M}_n$ is one-to-one, and hence isometric, by Corollary 32 and Lemma 26 (note that $\pi(\mathfrak{M}_n)$ is finite-dimensional, therefore a Banach algebra). Thus π is isometric on the norm dense *-subalgebra $\cup_{n=1}^{\infty} \mathfrak{M}_n$, hence (by continuity) isometric everywhere. Therefore $\mathcal{J} = 0$.

Finally, there are no nontrivial ideals at all, since the closure of a proper ideal is proper.

Exercise. A von Neumann factor of finite type is type I if and only if it is finite dimensional. Hence, an infinite dimensional factor of finite type must be of type II_1 (*Hints.* If $\mathfrak{A}$ is a type I finite factor, use the Type I Structure Theorem of §6, together with the matrix construction in Appendix A, to show that $\mathfrak{A}$ is *-isomorphic to $M_n(\mathbf{C})$, for some $n < \infty$. For the converse, use the Wedderburn Structure Theorem. The last statement is a consequence of the first and the type classification of Theorem 3).

The remainder of this section is devoted to the examination of two particular hyperfinite von Neumann factors.

Example 1 (A hyperfinite factor of type II_1 [6]). Let $\mathfrak{A}$ be a UHF algebra of type $\{2^n\}$, τ its trace, and let $\pi: \mathfrak{A} \to \mathcal{B}(\hbar)$ be the cyclic representation constructed from the state τ (see the discussion preceding Lemma 25). Let $x_0 = I + \mathcal{L}$ be the cyclic vector associated with π ($\mathcal{L}$ = the left kernel of τ).

We now begin a list of facts about this representation.

(1) $\mathcal{L} = 0$ (*Proof.* Take $L \in \mathcal{L}$, $A \in \mathfrak{A}$. Then $[LA, LA]_\tau = \tau(A^*L^*LA) = \tau(AA^*L^*L) = [L, LAA^*]_\tau \leqslant \tau(L^*L) \cdot \tau(AA^*L^*LAA^*) = 0$, using first the trace property (Remark after Corollary 31), then the Schwarz Inequality (Lemma 18). Hence $\mathcal{L}$ is an ideal, and since $\mathcal{L} \neq \mathfrak{A}$ ($\tau(I) = 1$), $\mathcal{L} = 0$ by Theorem 21).

(2) x_0 is a cyclic trace vector for $\pi(\mathfrak{A})$ (*Proof.* By (1), $x_0 = I$, and for $A, B \in \mathfrak{A}$, $[\pi(A)\pi(B)x_0, x_0]_\tau = [AB, I]_\tau = \tau(AB) = \tau(BA) = [BA, I]_\tau = [\pi(B)\pi(A)x_0, x_0]_\tau$).

(3) π is a faithful representation (*Proof.* $\pi^{-1}(0)$ is an ideal $\neq \mathfrak{A}$, hence $\pi^{-1}(0) = 0$, by Theorem 21).

(4) x_0 is a cyclic trace vector for both $(\pi(\mathfrak{A}))''$ and $(\pi(\mathfrak{A}))'$ (*Proof.* Let $S', T' \in (\pi(\mathfrak{A}))'$. Then there exist sequences $A_n, B_n \in \mathfrak{A}$ with $S'x_0 = \lim_n \pi(A_n)x_0$ and $T'x_0 = \lim_n \pi(B_n)x_0$. Also

$[S'T'x_0, x_0]_\tau = \lim_{m,n}[\pi(A_m)\pi(B_n)x_0, x_0]_\tau = \lim_{m,n}[\pi(B_n)\pi(A_m)x_0, x_0]_\tau = [T'S'x_0, x_0]_\tau$. Clearly the same argument applies to S'', $T'' \in (\pi(\mathfrak{A}))''$, so x_0 is a trace vector for both algebras. Since x_0 is cyclic for $(\pi(\mathfrak{A}))''$, it is separating for this algebra (see the exercise after Proposition (5); hence is cyclic for $(\pi(\mathfrak{A}))'$ by Proposition 5(4)).

(5) $(\pi(\mathfrak{A}))''$ and $(\pi(\mathfrak{A}))'$ are hyperfinite II_1-factors (*Proof.* Let $\mathcal{R} = (\pi(\mathfrak{A}))''$, and let G be a central projection in $\mathcal{R}$. For $T \in \mathcal{R}$, let

$$f(T) = \tau(TG),$$

and

$$g(T) = \tau(T(I - G)).$$

Since $\tau = \omega_{x_0}|\mathcal{R}$, $f = \omega_{Gx_0}|\mathcal{R}$ and $g = \omega_{(I-G)x_0}|\mathcal{R}$. Thus f and g are weakly continuous positive linear functionals on $\mathcal{R}$ which vanish on all operators of the form $ST - TS$, with S, $T \in \mathcal{R}$.

Since we are assuming that $\mathfrak{A}$ is a UHF algebra of type $\{2^n\}$, $\mathfrak{A}$ contains an increasing sequence $\{\mathfrak{M}_n\}$ of factors (with $\mathfrak{M}_n$ of type I_{2^n}) whose union is norm dense in $\mathfrak{A}$. It follows immediately that if we identify $\mathfrak{A}$ with its image $\pi(\mathfrak{A})$, the union $\mathfrak{M} = \bigcup_{n=1}^\infty \mathfrak{M}_n$ is strongly dense in $\mathcal{R}$.

By uniqueness of the normalized trace tr_n on $\mathfrak{M}_n$ (remark before Lemma 24), we see that $f|\mathfrak{M}_n = \tau(G)\cdot tr_n$ and $g|\mathfrak{M}_n = (1 - \tau(G))\cdot tr_n$. Further, $f|\mathfrak{M} = \tau(G)\cdot(\tau|\mathfrak{M})$ and $g|\mathfrak{M} = (1 - \tau(G))\cdot(\tau|\mathfrak{M})$. By weak continuity of f and g, $f = \tau(G)\cdot\tau$ and $g = (1 - \tau(G))\cdot\tau$. By theorem 6(iv), $\tau(G) \neq 0$ if $G \neq 0$. But then $g = (\tau(G))^{-1}(1 - \tau(G))\cdot f$, and since $g(I - G) = 0$, $\tau(G) = 1$. Thus $\tau(I - G) = 0$, and by Theorem 6(iv) again, $G = I$.

Therefore $\mathcal{R}$ is a factor of finite type (Theorem 6) which is infinite dimensional. By the exercise before this example, $\mathcal{R}$ is a factor of type II_1. By Corollary 6, $\mathcal{R}' = (\pi(\mathfrak{A}))'$ is *-antiisomorphic with $\mathcal{R}$, and hence is also of type II_1).

(6) Both $(\pi(\mathfrak{A}))''$ and $(\pi(\mathfrak{A}))'$ are singly generated as von Neumann algebras (*Proof.* They are both hyperfinite and Corollary 31 applies).

(7) The Hilbert space $\mathfrak{h}$ is separable (*Proof.* For $A \in \mathfrak{A}$, let $||A||_2 = \tau(A^*A)^{1/2}$. Then $||\cdot||_2$ is a norm on $\mathfrak{A}$ by Theorem 6(iv). Since $0 \leqslant A^*A \leqslant ||A||^2 \cdot I$ (§0, 26) and 27)), $||A||_2 \leqslant ||A||$, for each $A \in \mathfrak{A}$. Hence a norm dense linear subspace of $\mathfrak{A}$ is $||\cdot||_2$-dense in $\mathfrak{h}$. But any UHF algebra is norm separable—a countable dense set can be obtained, for example, by taking linear combinations, with Gaussian numbers as coefficients, of the semigroup generated by the identity, a generator (Theorem 20) and its adjoint).

Before presenting the next example, we develop a fragment of representation theory. We consider an arbitrary C^*-algebra $\mathfrak{A}$ with $I \in \mathfrak{A}$, a state f of $\mathfrak{A}$, and the cyclic $*$-representation $\pi \colon \mathfrak{A} \to \mathcal{B}(\mathfrak{h})$ defined by f, with cyclic vector $x_0 = I + \mathfrak{L}$, where $\mathfrak{L}$ is the left kernel of f.

The set of all states of $\mathfrak{A}$ is easily seen to be a convex subset of the unit ball (Corollary 16) of the conjugate space $\mathfrak{A}^*$. For each $A \in \mathfrak{A}$, the maps φ_A defined on $\mathfrak{A}^*$ by the formula $\varphi_A(f) = |f(A^*A)| - f(A^*A)$, and the map ϵ defined by $\epsilon(f) = 1 - f(I)$, are continuous in the weak $*$ topology on $\mathfrak{A}^*$. But the set of states is describable as the intersection

$$\epsilon^{-1}(0) \cap (\cap \{\varphi_A^{-1}(0) \colon A \in \mathfrak{A}\}),$$

and hence is a closed set in the weak $*$ topology of $\mathfrak{A}^*$.

A *pure state* is an extreme point of the convex set of states. The above discussion implies, in conjunction with the Krein-Milman Theorem, that every C^*-algebra has an abundance of pure states.

Suppose now that f is a pure state of $\mathfrak{A}$, and that g is a positive linear functional on $\mathfrak{A}$ with $0 \leqslant g \leqslant f$. We assert that $g = \lambda \cdot f$, for some λ with $0 \leqslant \lambda \leqslant 1$. To see this, put $\lambda = g(I)$, and observe that if $\lambda = 0$, then $g = 0$, since $g \geqslant 0$. Likewise, if $g(I) = 1$, then $(f - g)(I) = 0$ and $f = g$, since $f - g \geqslant 0$. Finally, if $0 < \lambda < 1$, write

$$f = \lambda(\lambda^{-1} \cdot g) + (1 - \lambda)[(1 - \lambda)^{-1} \cdot (f - g)].$$

By purity of f, $g = \lambda \cdot f$, as required.

Exercise. If the only positive linear functionals dominated by a state f are multiples of f, then f is pure.

Next we show that the cyclic $*$-representation π associated with a pure state f is irreducible. To do this, let E be a projection in $(\pi(\mathfrak{A}))'$. Define $g(A) = [\pi(A)Ex_0, x_0]_f$. Then if $A \in \mathfrak{A}^+$, $\pi(A) \in \pi(\mathfrak{A})^+$ and $0 \leqslant E\pi(A)E = \pi(A)E \leqslant \pi(A)$, so $0 \leqslant g \leqslant f$. Thus $g = g(I) \cdot f$ by the above, i.e., $[\pi(A)Ex_0, x_0]_f = [Ex_0, x_0]_f \cdot [\pi(A)x_0, x_0]_f$. Since x_0 is cyclic for $\pi(\mathfrak{A})$ and $[\pi(A)Ex_0, x_0]_f = [\pi(A)x_0, Ex_0]_f$, we can choose a sequence $A_n \in \mathfrak{A}$ so that $Ex_0 = \lim_n \pi(A_n)x_0$, and then $[Ex_0, x_0]_f^2 = \lim_n [Ex_0, x_0]_f \cdot [\pi(A_n)x_0, x_0]_f = \lim_n [\pi(A_n)Ex_0, x_0]_f = [Ex_0, x_0]_f$. Thus $||Ex_0||^2 = [Ex_0, x_0]_f = 0$ or 1. If $||Ex_0|| = 0$, $0 = Ex_0 = \pi(\mathfrak{A})Ex_0 = E\pi(\mathfrak{A})x_0$, so $E = 0$. If $||Ex_0|| = 1$, then $||(I - E)x_0||^2 = [(I - E)x_0, x_0]_f = 0$, and as before, $E = I$. Hence $\pi(\mathfrak{A})$ acts irreducibly on $\mathfrak{h}$ and $\pi(\mathfrak{A})'' = \mathcal{B}(\mathfrak{h})$.

Example 2 (A hyperfinite factor of type I_∞). We show that $\mathcal{B}(\mathcal{H})$, with $\mathcal{H}$ separable, is a hyperfinite von Neumann factor. Let $\mathfrak{A}$ be a UHF algebra. By the above discussion, $\mathfrak{A}$ has pure states. If f is one such, the (nonzero) cyclic $*$-representation π arising therefrom is irreducible, by the foregoing (moreover, any nonzero representation of a UHF algebra is necessarily faithful by Theorem 21). As in part (7) of Example 1, every such cyclic $*$-representation of a UHF algebra is on a separable Hilbert space $\mathfrak{h}$. Thus $\mathcal{B}(\mathfrak{h}) = \pi(\mathfrak{A})''$ is hyperfinite, and since $\mathfrak{h} \cong \mathcal{H}$, $\mathcal{B}(\mathcal{H}) \cong \mathcal{B}(\mathfrak{h})$ is too.

The representation theory of UHF algebras is quite rich. In particular, there are uncountably many (faithful) *nonisomorphic* *-representations π such that $\pi(\mathcal{A})''$ is a factor of type III ([17]).

Exercise. Let $\mathcal{A}$ be a factor of type II_1 (e.g., the one in Example 1 above). Show that $M_\infty(\mathcal{A})$ is a factor of type II_∞.

It is known that every UHF algebra has factor representations of type II_∞ (there seems to be no convenient reference to this fact in the literature). The following sketchy construction contains the basic ideas for a proof. Let $\mathcal{A}$ be a UHF algebra of type $\{2^n\}$, let π be an irreducible representation (as in Example 2 above), and let σ be a type II_1 representation (as in Example 1 above). Then the tensor product representation $\pi \otimes \sigma$ of the (spatial) C^*-tensor product $\mathcal{A} \otimes \mathcal{A}$ (a UHF algebra of type $\{2^{2n}\}$) is a factor representation of type II_∞ (roughly speaking, $I_\infty \otimes II_1 = II_\infty$).

APPENDIX A

Direct Sums and Matrix Algebras

Direct Sums. Let $\{\mathcal{Q}_i\}_{i \in I}$ be a family of von Neumann algebras, and suppose that $\mathcal{Q}_i$ acts on the Hilbert space $\mathcal{H}_i$, for each $i \in I$. The *Hilbert space direct sum* $\mathcal{H} = \Sigma_{i \in I} \oplus \mathcal{H}_i$ is the set of all mappings $x: I \to \cup_{i \in I} \mathcal{H}_i$ such that
 (i) $x_i (= x(i)) \in \mathcal{H}_i$, for each $i \in I$, and
 (ii) $\Sigma_{i \in I} ||x_i||^2 < \infty$
(here $\Sigma_{i \in I}||x_i||^2 = \text{LUB}_F\{\Sigma_{i \in F}||x_i||^2: F \text{ finite} \subset I\}$—the condition $\Sigma_{i \in I}||x_i||^2 < \infty$ implies $x_i = 0$ for all but a *countable* set of indices $i \in I$—see [7; Theorems 1 and 2, p.19]). We furnish $\mathcal{H} = \Sigma_{i \in I} \oplus \mathcal{H}_i$ with the *inner product*

$$((x_i)|(y_i)) = \Sigma_{i \in I}(x_i|y_i),$$

where $(x_i), (y_i) \in \mathcal{H}$, and the linear operations are defined co-ordinatewise:

$$a(x_i) + (y_i) = (ax_i + y_i) \qquad (a \text{ complex})$$

(Note that the *norm* in $\mathcal{H}$ is given by $||(x_i)|| = (\Sigma_{i \in I}||x_i||^2)^{1/2}$. Also, in any inner product space, $2|(x|y)| \leq ||x||^2 + ||y||^2$. Thus the inner product defined above is finite, since $|\Sigma_{i \in I}(x_i|y_i)| \leq \Sigma_{i \in I}|(x_i|y_i)| \leq \frac{1}{2}\Sigma_{i \in I}(||x_i||^2 + ||y_i||^2) = \frac{1}{2}\Sigma_{i \in I}||x_i||^2 + \frac{1}{2}\Sigma_{i \in I}||y_i||^2 < \infty$). Furthermore, $\mathcal{H}$ is complete (imitate the proof for $l^2(I)$).

Given a family $(T_i)_{i \in I}$ of operators, with each $T_i \in \mathcal{B}(\mathcal{H}_i)$, such that $\sup_{i \in I}||T_i|| < \infty$, the formula

$$T(x_i) = (T_i x_i)$$

defines a bounded operator on $\mathcal{H}$, since $||T(x_i)||^2 = \Sigma_{i \in I}||T_i x_i||^2 \leq \Sigma_{i \in I}||T_i||^2 \cdot ||x_i||^2 \leq \sup_{i \in I}||T_i||^2 \cdot \Sigma_{i \in I}||x_i||^2 < \infty$. Let $\mathcal{A} = \{A = (A_i) : A_i \in \mathcal{A}_i, \sup_{i \in I}||A_i|| < \infty\}$ (the linear operations, adjoint and product are co-ordinatewise).

ASSERTION. $\mathcal{A}$ *is a von Neumann algebra, called the (external) direct sum of the family* $\{\mathcal{A}_i\}_{i \in I}$, *written*

$$\mathcal{A} = \Sigma_{i \in I} \oplus \mathcal{A}_i.$$

Proof. Let $E_i \in \mathcal{B}(\mathcal{H})$ be the projection onto $\mathcal{H}_i$ and let $\mathcal{B} = \{B = (B_i) : B_i \in \mathcal{A}_i', \sup_{i \in I}||B_i|| < \infty\}$. Both $\mathcal{A}$ and $\mathcal{B}$ are *-algebras of operators on $\mathcal{H} = \Sigma_{i \in I} \oplus \mathcal{H}_i$, and $AB = BA$, for each $A \in \mathcal{A}$ and $B \in \mathcal{B}$ (i.e., $\mathcal{A} \subset \mathcal{B}'$ or equivalently, $\mathcal{B} \subset \mathcal{A}'$).

Now suppose $T \in \mathcal{B}(\mathcal{H})$ commutes with each operator in $\mathcal{B}$. Since

$$(E_i)_j = \begin{cases} I_{\mathcal{H}_i} & \text{if } i = j \\ 0 & \text{if } i \neq j \end{cases}$$

we clearly have $E_i \in \mathcal{B}$, so $TE_i = E_i T$, for each $i \in I$, and if $T_i = T|_{\mathcal{H}_i}$, then $T_i \in \mathcal{B}(\mathcal{H}_i)$ and $T = (T_i)$. Hence, if $B \in \mathcal{B}$, $B = (B_i)$, with each $B_i \in \mathcal{A}_i'$, then $(T_i B_i) = TB = BT = (B_i T_i)$, so $T_i B_i = B_i T_i$, for each $B_i \in \mathcal{A}_i'$, i.e., $T_i \in \mathcal{A}_i'' = \mathcal{A}_i$ (by the Double Commutant Theorem). Thus $T \in \mathcal{A}$, so $\mathcal{B}' = \mathcal{A}$, proving the assertion. Interchanging the roles of $\mathcal{A}_i$ and $\mathcal{A}_i'$, we obtain $\mathcal{A}' = \mathcal{B}$, proving the following result.

COROLLARY. *If* $\mathcal{A} = \Sigma_{i \in I} \oplus \mathcal{A}_i$, *then* $\mathcal{A}' = \Sigma_{i \in I} \oplus \mathcal{A}_i'$, *for any family* $\{\mathcal{A}_i\}_{i \in I}$ *of von Neumann algebras.*

Exercise. Let $\mathcal{A}$ be a von Neumann algebra acting on the Hilbert space $\mathcal{H}$, and let $\{E_i\}_{i \in I}$ be a family of central projections in $\mathcal{A}$ which are orthogonal in pairs, and whose LUB is I. In this case, we call $\mathcal{A}$ the (internal) *direct sum* of the ideals $\mathcal{A}E_i$. Let $\mathcal{H}_i = \operatorname{ran}(E_i)$. Then

$$\mathcal{A} \cong \Sigma_{i \in I} \oplus \mathcal{A}_i \qquad (\text{on } \Sigma_{i \in I} \oplus \mathcal{H}_i)$$

where $\mathcal{A}_i = \mathcal{A}|_{\mathcal{H}_i}$ and $\cong$ denotes a spatial *-algebra isomorphism which is implemented by an isometry

$$V: \mathcal{H} = \Sigma_{i \in I} \oplus \mathcal{H}_i,$$

so that an operator $A \in \mathcal{A}$ is mapped to VAV^{-1}.

Matrix Algebras. Let $n (\geqslant 2)$ be any cardinal number, and let $\{E_i\}_{i=1}^n$ be a family of orthogonal projections in a von Neumann algebra $\mathcal{A}$, with LUB $E_i = I$. Suppose further that there are partial isometries $U_i \in \mathcal{A}$ with $U_i^* U_i = E_1$ and $U_i U_i^* = E_i$. If $\mathcal{A}$ acts on $\mathcal{H}$ and if $\mathcal{H}_1 = \operatorname{ran}(E_1)$, then there is a spatial *-isomorphism between $\mathcal{A}$ (on $\mathcal{H}$) and the algebra $M_n(E_1 \mathcal{A} E_1)$ (acting on $\Sigma_{i=1}^n \oplus \mathcal{H}_1 = $ the direct sum of n copies of $\mathcal{H}_1$) of all $n \times n$ matrices with entries from $E_1 \mathcal{A} E_1$ which act as bounded operators on $\Sigma_{i=1}^n \oplus \mathcal{H}_1$.

(Note that if $n < \infty$, *all* $n \times n$ matrices act as bounded operators, but this is not the case if n is infinite, e.g., if $n = \aleph_0$, the matrix

$$\begin{pmatrix} I & 0 & 0 & . & . & . \\ I & 0 & 0 & . & . & . \\ I & 0 & 0 & . & . & . \\ . & . & . & & & \\ . & . & . & & & \\ . & . & . & & & \end{pmatrix}$$

does not act as a bounded operator.)

This isomorphism is given by

$$A \mapsto (U_i^* A U_j)_{i,j=1}^n.$$

Note that since $E_1 U_i^* = U_i^* U_i U_i^* = U_i^*$ and $U_j E_1 = U_j U_j^* U_j = U_j$, we have $U_i^* A U_j = E_1(U_i^* A U_j) E_1 \in E_1 \mathfrak{A} E_1$.

ASSERTION. *Let $\mathfrak{A}$ be a n-homogeneous von Neumann algebra and let $\mathfrak{Z}$ be its center. Then there is a spatial $*$-isomorphism*

$$\mathfrak{A} \cong M_n(\mathcal{C}),$$

where $\mathcal{C}$ is an abelian von Neumann algebra $*$-isomorphic to $\mathfrak{Z}$.

Proof. Choose an orthogonal family $\{E_i\}_{i=1}^n$ of equivalent abelian projections with LUB I, and partial isometries $\{U_i\}_{i=1}^n$ in $\mathfrak{A}$ with $U_i^* U_i = E_1$ and $U_i U_i^* = E_i$. Taking $\mathcal{C} = E_1 \mathfrak{A} E_1 = \mathfrak{Z} E_1$ (Corollary 8), we obtain the desired spatial $*$-isomorphism, since faithfulness of E_1 (see Exercise (2) after Theorem 3 and the remark after Lemma 14) forces the map $Z \mapsto Z E_1$ of $\mathfrak{Z}$ onto $\mathfrak{Z} E_1$ to be one-to-one, hence a $*$-isomorphism (nonspatial if $n \geqslant 2$).

Exercises. (1) Let $\{E_i\}_{i=1}^n$ be an orthogonal family of projections (n = any cardinal number) in a von Neumann algebra $\mathfrak{A}$ with $I = \text{LUB } E_i$. Suppose $\{U_i\}_{i=1}^n$ is a family of partial isometries in $\mathfrak{A}$ with $U_i^* U_i = E_1$ and $U_i U_i^* = E_i$. Define $E_{ij} = U_i U_j^*$. Show that the E_{ij}'s satisfy the usual rule $E_{ij} E_{kl} = \delta_{jk} E_{il}$ for matrix units (*Note.* $E_{ii} = E_i$).

Next, let $\mathfrak{D} = \{E_{ij}\}' \cap \mathfrak{A}$. Prove that the mapping $D \mapsto D E_i$ is a $*$-algebra isomorphism of $\mathfrak{D}$ with $E_i \mathfrak{A} E_i$ (*Hint.* If $D \in \mathfrak{D}$ annihilates any E_{ij}, then $D = 0$).

(2) Show that $\mathfrak{D} = \mathfrak{A} \cap \mathfrak{A}' = \mathfrak{Z}$ (the center of $\mathfrak{A}$) (*Hint.* Show that $\mathfrak{D} E_i = \mathfrak{Z} E_i$, then observe that the representation of an operator $A = D E_i \in \mathfrak{D} E_i$ is unique).

DIRECT SUMS AND MATRIX ALGEBRAS 99

(3) Prove that $E_i \mathcal{A} E_j = \mathcal{D} E_{ij}$, and that the representation of an operator in $E_i \mathcal{A} E_j$ in the form $D E_{ij}$ is unique (*Hint*. Use the hint in exercise (1) above).

(4) Let n be finite. Then each $A \in \mathcal{A}$ can be written uniquely as

$$A = \sum_{i,j=1}^{n} D_{ij} E_{ij},$$

where $D_{ij} \in \mathcal{D}$. In fact, $D_{ij} = \sum_{k=1}^{n} E_{ki} A E_{jk}$. Also $D_{ij} E_{ij} = E_i A E_j$.

(5) Let $\mathcal{A}$ be a von Neumann algebra acting on the Hilbert space $\mathcal{H}$, and let $n \,(\geqslant 2)$ be any cardinal number. Show that the n-fold direct sum $\sum_{i=1}^{n} \oplus \mathcal{A}$ of $\mathcal{A}$ with itself (acting on $\sum_{i=1}^{n} \oplus \mathcal{H}$) is spatially *-isomorphic with the (von Neumann) subalgebra of $M_n(\mathcal{A})$ consisting of all diagonal matrices.

(6) Let $\mathcal{A}$ be a von Neumann algebra acting on the Hilbert space $\mathcal{H}$, and let $n \,(\geqslant 2)$ be any cardinal number. Denote by $n\mathcal{A}$ the *-subalgebra of $M_n(\mathcal{A})$ consisting of all matrices of the form

$$\begin{pmatrix} A & & & & \\ & A & & 0 & \\ & & A & & \\ & & & \cdot & \\ & 0 & & & \cdot \end{pmatrix},$$

where $A \in \mathcal{A}$. Show that $(n\mathcal{A})' = M_n(\mathcal{A}')$ and also that $n\mathcal{A} = M_n(\mathcal{A}')'$.

APPENDIX B

The Coupling Theorem

One of the more profound aspects of the spatial theory of von Neumann algebras is the duality that exists between an algebra and its commutant. A fragment of this duality was expounded in §5, where it was shown that 'type I' is preserved in passing from an algebra to its commutant (Theorem 7). Our eventual aim is to show that the remaining types (II and III) are preserved in this passage. The proof requires ideas and techniques not encountered in the type I setting, and in this appendix we develop the basic machinery. We begin by analyzing the range of a cyclic projection.

THE 'B − T THEOREM'. *Let $\mathcal{A}$ be a von Neumann algebra acting on the Hilbert space $\mathcal{H}$, and let $a \in \mathcal{H}$. If $b \in [\mathcal{A}a]$, then we can find operators $B, H \in \mathcal{A}$ with $H \geqslant 0$, and a vector $c \in \text{null}(H)^{\perp}$ so that $b = Bc$ and $a = Hc$.*

Remark. If $T = H^{-1}$ (possibly unbounded), then $b = BTa$, hence the designation '$B - T$'.

Proof. We first describe a general construction due to Friedrichs and modified by Murray and von Neumann for the present purpose [13; Lemmas 9.1.1 to 9.1.5]. Given a sequence of operators $T_n \in \mathcal{A}$, put

$$\mathcal{D} = \left\{ x \in \mathcal{H}: \sum_{n=1}^{\infty} ||T_n x||^2 < \infty \right\}.$$

From the inequality $||T(x+y)||^2 \leq 2(||Tx||^2 + ||Ty||^2)$, we see that $\mathfrak{D}$ is a linear subspace of $\mathcal{H}$. The linear map $T: \mathfrak{D} \to \Sigma_{n=1}^{\infty} \oplus \mathcal{H}$ defined by

$$Tx = (T_1 x, T_2 x, \ldots)$$

gives rise to an inner product structure on $\mathfrak{D}$ if, for $x, y \in \mathfrak{D}$, we define

$$(x|y)_{\mathfrak{D}} = (x|y) + (Tx|Ty)$$

$$\left(= (x|y) + \sum_{n=1}^{\infty} (T_n x | T_n y) \right).$$

Now $||x||_{\mathfrak{D}}^2 = ||x||^2 + ||Tx||^2 \geq ||x||^2$, for all $x \in \mathfrak{D}$. We assert that T has closed graph. To see this, let $||x - x_n|| \to 0$ and $||y - Tx_n|| \to 0$, where $x_n \in \mathfrak{D}$. We must show that $x \in \mathfrak{D}$ and $Tx = y$. Given $\epsilon > 0$, choose k large enough so that $||x_m - x_n|| \leq \sqrt{\epsilon}/2$ and $||T(x_m - x_n)|| < \sqrt{\epsilon}/2$, whenever $m, n > k$. Then

$$||x_m - x_n||^2 + \sum_{i=1}^{\infty} ||T_i(x_m - x_n)||^2 \leq \epsilon.$$

Letting $m \to \infty$, we see that

$$||x - x_n||^2 + \sum_{i=1}^{\infty} ||T_i(x - x_n)||^2 \leq \epsilon$$

by continuity of the norm and of the T's. In particular, $\Sigma_{i=1}^{\infty} ||T_i(x - x_n)||^2 < \infty$, so $x - x_n \in \mathfrak{D}$ and $x \in \mathfrak{D}$. Thus $||x - x_n||_{\mathfrak{D}}^2 \leq \epsilon$, so that $||x - x_n||_{\mathfrak{D}} \to 0$ as $n \to \infty$. But $||Tx - Tx_n|| \leq ||x - x_n||_{\mathfrak{D}} \to 0$, so $Tx = y$.

Now it is easy to see that completeness of the inner product space $(\mathfrak{D}, ||\cdot||_{\mathfrak{D}})$ is equivalent to T having closed graph. Hence $\mathfrak{D}$, under the norm $||\cdot||_{\mathfrak{D}}$ is a Hilbert space. We now regard the original inner product (on $\mathcal{H}$) as a bilinear form on $\mathcal{D}$. Since

$$||x|| \leq ||x||_{\mathcal{D}} \qquad (x \in \mathcal{D}),$$

this form is bounded (by 1). By [7; Theorem 1, p.38], there is a unique positive one-to-one operator D (of $||\cdot||_{\mathcal{D}}$-norm ≤ 1) on $\mathcal{D}$ with

$$(x|y) = (Dx|y)_{\mathcal{D}} \qquad (x, y \in \mathcal{D}).$$

Thus $0 \leq (Dx|x) = (D^2x|x)_{\mathcal{D}} \leq (Dx|x)_{\mathcal{D}} = (x|x)$, for each $x \in \mathcal{D}$. Hence D extends by continuity (relative to the $||\cdot||$-norm) to a unique operator, again denoted by D, on the closure $[\mathcal{D}]$ of $\mathcal{D}$. Clearly $\mathcal{D}$ is invariant under $\mathcal{C}'$ (as is $[\mathcal{D}]$), since it was defined using operators from $\mathcal{C}$ only. Further, if $U' \in \mathcal{C}'$ is unitary, the definition of $(\cdot|\cdot)_{\mathcal{D}}$ makes it clear that $(U'x|U'y)_{\mathcal{D}} = (x|y)_{\mathcal{D}}$, for all $x, y \in \mathcal{D}$. Thus $(DU'x|y)_{\mathcal{D}} = (U'x|y) = (x|U'^*y) = (Dx|U'^*y)_{\mathcal{D}} = (U'Dx|U'U'^*y)_{\mathcal{D}} = (U'Dx|y)_{\mathcal{D}}$, for all $x, y \in \mathcal{D}$, so that $DU'x = U'Dx$, for all $x \in [\mathcal{D}]$ (by continuity). Finally, if we extend D to all of $\mathcal{H}$ by putting $D = 0$ on $\mathcal{D}^{\perp}$, we have $DU' = U'D$, for all unitaries $U' \in \mathcal{C}'$, so that $D \in \mathcal{C}$ by remark (2) (iii) after Proposition 2. We have $0 \leq D \leq I$, and we can therefore form $H = D^{1/2} \in \mathcal{C}$. Now null$(D) = $ null(H) ($\S0$, 3)). If $Dx = 0$, then $(x|y) = (Dx|y)_{\mathcal{D}} = 0$, for all $y \in \mathcal{D}$, so that $x \in \mathcal{D}^{\perp}$. Conversely, if $(x|y) = 0$, for all $y \in \mathcal{D}$, then $(Dx|y)_{\mathcal{D}} = (x|y) = 0$, so $Dx = 0$. Thus null$(D) = \mathcal{D}^{\perp}$, so $[\mathcal{D}] = [\text{ran}(D)] = [\text{ran}(H)]$ ($\S0$, 2)).

We assert that $\mathcal{D} = $ ran(H). To see this, take $y \in [\mathcal{D}]$ and choose $y_n = Hx_n \in $ ran(H) with $||y - y_n|| \to 0$. An easy computation reveals that $||Dx_m - Dx_n||_{\mathcal{D}} = ||y_m - y_n|| \to 0$, so by completeness of $(\mathcal{D}, ||\cdot||_{\mathcal{D}})$, we can find $z \in \mathcal{D}$ with $||z - Dx_n||_{\mathcal{D}} \to 0$ (and also $||z - Dx_n|| \to 0$, because $||\cdot|| \leq ||\cdot||_{\mathcal{D}}$). Since $Dx_n = Hy_n$, $Hy = z \in \mathcal{D}$, and therefore ran$(H) \subset \mathcal{D}$. Also, because $||Hy - Dx_n||_{\mathcal{D}} \to 0$, we have $||Hy||_{\mathcal{D}}^2 = (Hy|Hy)_{\mathcal{D}} = \lim(Dx_n|Dx_n)_{\mathcal{D}} = \lim(x_n|Dx_n) = \lim||Hx_n||^2 = ||y||^2$, so that $||Hy||_{\mathcal{D}} = ||y||$, for all $y \in [\mathcal{D}]$. This proves that $[\mathfrak{D}]$ (under $||\cdot||$) is isometric with ran(H) (under $||\cdot||_{\mathfrak{H}}$), so that ran(H) is complete in the $||\cdot||_{\mathfrak{H}}$-norm. Now if $y \in \mathfrak{D}$, and $(Hx|y)_{\mathfrak{H}} = 0$, for all $x \in \mathcal{H}$, then certainly $(Dx|y)_{\mathfrak{H}} = 0$, because ran($D$) $\subset$ ran(H). But then $(x|y) = (Dx|y)_{\mathfrak{H}} = 0$, for all $x \in \mathcal{H}$, so $y = 0$. This shows that ran(H) = $\mathfrak{D}$, as asserted.

Turning now to the proof of the theorem, suppose $b \in [\mathfrak{A}a]$ and choose a sequence of operators $A_n \in \mathfrak{A}$ so that $||b - A_n a|| \leq 4^{-n}$. Put $T_n = 2^n(A_{n+1} - A_n)$, and use the T_n's to construct $\mathfrak{D}$, D and H as above. Then

$$||A_{n+1}a - A_n a||^2 \leq (||b - A_{n+1}a|| + ||b - A_n a||)^2$$
$$\leq (4^{-(n+1)} + 4^{-n})^2 \leq 2 \cdot 4^{-n},$$

so

$$\sum_{n=1}^{\infty} ||T_n a||^2 = \sum_{n=1}^{\infty} 4^n \cdot ||A_{n+1}a - A_n a||^2$$
$$\leq \sum_{n=1}^{\infty} 4^n (2 \cdot 4^{-n})^2 < \infty.$$

It follows that $a \in \mathfrak{D}$.

Since $||Hx||_{\mathfrak{H}} = ||x||$, if $x \in [\mathfrak{D}]$, and $Hx = 0$ if $x \in \mathfrak{D}^{\perp}$, we have $||Hx||_{\mathfrak{H}} \leq ||x||$, for all $x \in \mathcal{H}$. But then

$$||Hx||_{\mathfrak{H}}^2 = ||Hx||^2 + \sum_{n=1}^{\infty} ||2^n(A_{n+1} - A_n)Hx||^2$$
$$\geq 2^n \cdot ||(A_{n+1} - A_n)Hx||^2,$$

for every $n = 1, 2, \ldots$, so that

$$||(A_{n+1} - A_n)Hx|| \leq (2^{-n})^{1/2} \cdot ||x||,$$

and since this holds for all $x \in \mathcal{H}$,

$$||(A_{n+1} - A_n)H|| \leq (2^{-n})^{1/2}.$$

This proves that the sequence $\{A_n H\}$ is Cauchy in norm, and hence converges to an operator $B \in \mathfrak{A}$.

Finally, since $a \in \mathfrak{D} = \text{ran}(H)$, there is a vector $c \in \text{null}(H)^\perp$ with $Hc = a$. Also $Bc = (\text{norm-lim } A_n H)c = \lim(A_n Hc) = \lim A_n a = b$, as required, and the proof is complete.

Before applying the $B - T$ Theorem, we need an auxiliary result about cyclic projections.

LEMMA. *Let $\mathfrak{A}$ be a von Neumann algebra acting on the Hilbert space $\mathcal{H}$, let $A' \in \mathfrak{A}'$ and let $x, y \in \mathcal{H}$. If*

$E' =$ *the projection onto* $[\mathfrak{A}y]$, *and*

$F' =$ *the projection onto* $[\mathfrak{A}x]$,

and if $y = A'x$, then $E' \leqslant F'$. If, in addition, $x \in \text{null}(A')^\perp$, then $E' \sim F'$.

Proof. For $A \in \mathfrak{A}$, $Ay = AA'x = A'Ax$, so $[\mathfrak{A}y] = [\text{ran}(A'F')]$. But by Lemma 6, $\text{rp}(A'F') \sim \text{rp}(F'A'^*) \subset \text{ran}(F') = [\mathfrak{A}x]$, so $E' \leqslant F'$.

Now if $x \in \text{null}(A')^\perp$, $[\mathfrak{A}x] \subset \text{null}(A')^\perp = [\text{ran}(A'^*)]$, and this implies that $[\mathfrak{A}x] = [\text{ran}(F'A'^*)]$, so that $E' \sim F'$.

THE COUPLING THEOREM. *Let $\mathfrak{A}$ be a von Neumann algebra acting on $\mathcal{H}$, let $x, y \in \mathcal{H}$ and set*

$E =$ *the projection onto* $[\mathfrak{A}'x]$,

$F =$ *the projection onto* $[\mathfrak{A}'y]$,

$E' =$ *the projection onto* $[\mathfrak{A}x]$, *and*

$F' =$ *the projection onto* $[\mathfrak{A}y]$.

Assume $E \leqslant F$. Then $E' \leqslant F'$. Furthermore, $E \sim F$ if and only if $E' \sim F'$.

Note. $E, F \in \mathfrak{A}$, while $E', F' \in \mathfrak{A}'$.

Proof. By assumption, there is a partial isometry $U \in \mathfrak{A}$ with

$E = U^*U$ and $UU^* \leqslant F$. Let $w = Ux$. Then $w \in \operatorname{ran}(U) \subset [\mathcal{C}'y]$ with $x = U^*w$. By the $B - T$ Theorem, there are operators B', $H' \in \mathcal{C}'$ and a vector $z \in \operatorname{null}(H')^\perp$ so that $y = H'z$ and $w = B'z$. Then $[\mathcal{C}x] \subset [\mathcal{C}w]$, since $x = U^*w$. Let P and Q denote the projections onto $[\mathcal{C}w]$ and $[\mathcal{C}z]$, respectively. By the preceding lemma, $P \leqslant Q$, since $w = B'z$. Also $Q \sim F'$, since $y = H'z$ and $z \in \operatorname{null}(H')^\perp$. Thus, $E' \leqslant P \leqslant Q \sim F'$, and this implies that $E' \leqslant F'$, by the remark before Proposition 8. The last statement is an immediate consequence of the Schröder-Bernstein Theorem.

Remark. If E is the projection onto $[\mathcal{C}'x]$, and if $F \in \mathcal{C}$ with $F \leqslant E$, then $\operatorname{ran}(F) = [\mathcal{C}'Fx]$ (*Proof.* $\operatorname{ran}(F) = \operatorname{ran}(FE) = F(\operatorname{ran}(E)) = F[\mathcal{C}'x] = [F\mathcal{C}'x] = [\mathcal{C}'Fx]$).

APPENDIX C

Semifiniteness and the Commutant

We now extend some of the results of §5 to semifinite and purely infinite von Neumann algebra. The success of this program rests heavily on the Coupling Theorem (Appendix B), but before proceeding, we need to develop another piece of dimension theory.

In the list of facts that follows, $\mathcal{A}$ will denote a von Neumann algebra acting on the Hilbert space $\mathcal{H}$. We call a projection $E \in \mathcal{A}$ *properly infinite* if $E\mathcal{A}E$ is a properly infinite von Neumann algebra. If $\mathcal{A}$ is of type III, it is clear that every nonzero projection is properly infinite.

(1) A projection $E \in \mathcal{A}$ is properly infinite if and only if for each projection $G \in \mathcal{A}$ such that $EG \neq 0$, EG is infinite (*Proof.* Assume first that E has the last property. If $F \in E\mathcal{A}E$ were a finite central projection, then $F = C_F \cdot E$ by Corollary 7. By the remark before Corollary 11, F would be finite in $\mathcal{A}$ as well. But $F = E \cdot C_F$, if nonzero, would have to be infinite. Hence $F = 0$. Conversely, if E is properly infinite and $G \in \mathcal{A}$ is a central projection with $EG \neq 0$, then EG is central in $E\mathcal{A}E$, hence infinite there. By the remark before Corollary 11, EG is also infinite in $\mathcal{A}$).

(2) The central cover of a properly infinite projection is itself properly infinite (*Proof.* If E is a properly infinite projection and G is a central projection with $C_E \cdot G \neq 0$, then $EG \neq 0$, so EG is infinite by (1), as is therefore $C_E \cdot G \geq EG$).

We call a projection $E \in \mathcal{A}$ *countably decomposable* if every family $\{E_n\}$ of orthogonal projections in $\mathcal{A}$ with $E_n \leqslant E$ is at most countable (i.e., the algebra $E\mathcal{A}E$ is countably decomposable in the language of Theorem 16).

(3) If $E \in \mathcal{A}$ is a countably decomposable projection and $F \in \mathcal{A}$ is a properly infinite projection with $C_E \leqslant C_F$, then $E \lesssim F$ (*Proof.* Let $\{E_n\}$ be a maximal orthogonal family of non-zero projections with $E_n \leqslant E$ and $E_n \lesssim F$. Such E_n's exist by Lemma 8. Put $G = E - \text{LUB } E_n$. If $G \neq 0$, then $C_G \leqslant C_E \leqslant C_F$, so by Lemma 8 again, there are subprojections $0 \neq F_0 \leqslant F$ and $0 \neq G_0 \leqslant G$ in $\mathcal{A}$ with $F_0 \sim G_0$. But then $G_0 \leqslant E$ and $G_0 \perp E$, for all n, with $G_0 \lesssim F$, contradicting maximality of $\{E_n\}$. Thus $E = \text{LUB } E_n$. Since E is countably decomposable, $\{E_n\}_{n=1}^{\infty}$ is a countable family. Now write $F = \text{LUB } F_n$, where $\{F_n\}_{n=1}^{\infty}$ is a sequence of orthogonal projections in $\mathcal{A}$, with each $F_n \sim F$, as in the proof of Corollary 14, working first in the properly infinite algebra $F\mathcal{A}F$, then using the remark before Corollary 11. Then $E_n \lesssim F_n$, so Additivity of Equivalence gives $E = \text{LUB } E_n \lesssim \text{LUB } F_n = F$).

(4) Two countably decomposable, properly infinite projections with the same central cover are equivalent (*Proof.* This is immediate from (3) and the Schröder-Bernstein Theorem).

(5) If $\mathcal{A}$ is a type III von Neumann algebra, and if $E, F \in \mathcal{A}$ are countably decomposable projections, then $E \sim F$ if and only if $C_E = C_F$ (*Proof.* We may assume $E \neq 0 \neq F$. Then by (1) and the definition of 'type III', E and F are properly infinite. The result now follows from (4) and the remark after Lemma 14).

(6) If $\mathcal{A}$ is type III, and $0 \neq E \in \mathcal{A}$ is a projection, then $E\mathcal{A}E$ is also type III (*Proof.* If $F \in E\mathcal{A}E$ were a finite projection, F would also be finite in $\mathcal{A}$ by the remark before Corollary 11).

(7) Cyclic projections are countably decomposable (*Proof.* Let E be the (cyclic) projection onto $[\mathcal{A}'x]$, where $x \in \mathcal{H}$. If $\{E_n\}$ is an orthogonal family of projections in $\mathcal{A}$ with each $E_n \leqslant E$, then $\Sigma_n ||E_n x||^2 \leqslant ||Ex||^2 = ||x||^2 < \infty$, and hence at most a countable number of terms in the sum are nonzero. But if $E_n x = 0$, then $\{0\} = [\mathcal{A}'E_n x] = [E_n \mathcal{A}'x] = \mathrm{ran}(E_n E) = \mathrm{ran}(E_n)$, so the family $\{E_n\}$ contains at most a countable number of nonzero projections).

(8) For $x \in \mathcal{H}$, let E (resp., E') be the projection onto $[\mathcal{A}'x]$ (resp., $[\mathcal{A}x]$). Then $C_E = C_{E'}$. (*Proof.* By remark (3) after Corollary 4, $\mathrm{ran}(C_E) = [\mathcal{A}\mathcal{A}'\mathrm{ran}(E)] = [\mathcal{A}\mathcal{A}'x] = [\mathcal{A}'\mathcal{A}x] = [\mathcal{A}\mathcal{A}'\mathrm{ran}(E')] = \mathrm{ran}(C_{E'})$).

(9) If $\mathcal{A}$ is a finite von Neumann algebra with no type I summand, then $\mathcal{A}'$ is semifinite (*Proof.* Suppose $\mathcal{A}'$ had a type III summand. Then for some central projection $H \neq 0$, $\mathcal{A}'H$ is a type III algebra. Take $0 \neq x \in \mathrm{ran}(H)$ and consider the projection $E \in \mathcal{A}$ onto $[\mathcal{A}'x]$. By the Halving Lemma, we can write $E = F + G$, with $F \perp G$, $F \sim G$, so $C_F = C_G = C_E$, by Exercise (2) after Theorem 3 and the remark after Lemma 14. Hence E and F are inequivalent projections in $\mathcal{A}$ having the same central cover. Now $\mathrm{ran}(F) = [\mathcal{A}'Fx]$ by the remark after the Coupling Theorem (Appendix B). Let E' be the projection onto $[\mathcal{A}x]$, and F' the projection onto $[\mathcal{A}Fx]$. Then $E', F' \in \mathcal{A}'$ and $F' \leqslant E' \leqslant H$, so actually $E', F' \in \mathcal{A}'H$. By (7), E' and F' are countably decomposable, and by the remark before (1), E' and F' are properly infinite, being nonzero. Furthermore, E' and F' are inequivalent, by the Coupling Theorem. But by (8), $C_{E'} = C_E = C_F = C_{F'}$, which by (5) contradicts the supposition that $\mathcal{A}'H$ was type III. Thus, in fact, $\mathcal{A}'$ has no type III summand, hence is semifinite. Note that $\mathcal{A}'$ has no type I summand either, by Theorem 7 and the Corollary of Appendix A).

(10) If $\mathcal{A}$ is semifinite, so is $\mathcal{A}'$ (*Proof.* By Theorem 7 and the Corollary of Appendix A, we may assume that $\mathcal{A}$ and $\mathcal{A}'$ have no type I summands. Let F be a faithful finite projection in $\mathcal{A}$ (Proposition 4). Then $\mathcal{A}' \cong \mathcal{A}'F \cong (\mathcal{A}')_F = (\mathcal{A}_F)'$ by Lemma 13, the second exercise before Corollary 5, and by Corollary 5. Also $\mathcal{A}_F \cong F\mathcal{A}F$ (finite) by the exercise just cited. By (9) above, $\mathcal{A}'$ is semifinite).

(11) A von Neumann algebra is semifinite if and only if it is *-isomorphic to a von Neumann algebra with finite commutant (*Proof.* Imitate the proof of Theorem 8, replacing 'type I' by 'semifinite' and 'abelian' by 'finite', and use Proposition 4 together with (10) above).

(12) If $\mathcal{A}$ is of type I, II or III, then so is its commutant $\mathcal{A}'$ (*Proof.* This follows directly from the Corollary of Appendix A, Theorem 7 and (10) above).

References

1. DIXMIER, J. *Les Algèbres d'Operateurs dans l'Espace Hilbertien (Algèbres de von Neumann)*, Gauthier-Villars, Paris, Deuxième Édition, (1969).

2. DIXMIER, J. *Les C^*-Algèbres et leurs Représentations*, Gauthier-Villars, Paris (1964).

3. DOUGLAS, R. and PEARCY, C. 'Von Neumann algebras with a single generator,' *Michigan Math. J.* **16** 21–26 (1969).

4. DOUGLAS, R. and TOPPING, D. 'Operators whose squares are zero,' *Rev. Romaine Math. Pures Appl.* **12** 647–652 (1967).

5. GLIMM, J. 'On a certain class of operator algebras,' *Trans. Amer. Math. Soc.* **95** 318–340 (1960).

6. GLIMM, J. 'Type I C^*-algebras,' *Ann. Math.* **73**, 572–612 (1961).

7. HALMOS, P. *Introduction to Hilbert Space and the Theory of Spectral Multiplicity*, Chelsea Publishing Company, (1951).

8. HALMOS, P. *Finite-Dimensional Vector Spaces*, Second Edition, D. Van Nostrand (1958).

9. HALMOS, P. *A Hilbert Space Problem Book*, D. Van Nostrand, (1967).

10. KAPLANSKY, I. *Rings of Operators*, W. A. Benjamin, (1968).

11. LEBOW, A. 'A Schroeder-Bernstein theorem for projections,' *Proc. Amer. Math. Soc.* **19**, 144–145 (1968).

12. LOOMIS, L. *Abstract Harmonic Analysis*, D. Van Nostrand, (1953).

13. MURRAY, F. and VON NEUMANN, J. 'On rings of operators,' *Ann. of Math.* **37**, 116–229 (1936); see also: J. von Neumann, Collected Works, Vol. III, 6–119.

REFERENCES

14. VON NEUMANN, J. 'Zur Algebra der Funktionaloperatoren und Theorie der normalen Operatoren,' *Math. Ann.* **102**, 370-427 (1929); see also: Collected Works, Vol. II, 86-143.

15. VON NEUMANN, J. 'Über Funktionen von Funktionaloperatoren,' *Ann. of Math.* **32**, 191-226 (1931); see also: Collected Works, Vol. II, 177-212.

16. PEARCY, C. 'On certain von Neumann algebras which are generated by partial isometries,' *Proc. Amer. Math. Soc.* **15**, 393-395 (1964).

17. POWERS, R. 'Representations of uniformly hyperfinite algebras and their associated von Neumann rings,' *Ann. of Math.* **86**, 138-171 (1967).

18. RICKART, C. *General Theory of Banach Algebras*, D. Van Nostrand, (1960).

19. RIESZ, F. and SZ-NAGY, B. *Functional Analysis*, Ungar, (1955).

20. SAITÔ, T. 'Generators of certain von Neumann algebras,' *Tôhoku Math. J.* **20**, 101-105 (1968).

21. SCHWARTZ, J. *W*-Algebras*, Gordon and Breach, (1967).

22. TOPPING, D. 'UHF algebras are singly generated,' *Math. Scand.* **22** 224-226 (1968).

23. VIGIER, J.-P. *Étude sur les suites infinies d'operateurs hermitiens*, Thèse N° 1089, Geneve, (1946).

24. WOGEN, W. 'On generators for von Neumann algebras,' *Bull. Amer. Math. Soc.* **75**, 95-99 (1969).

QA
326
T66

FEB 2 1977